LIVING
IN
SPACE

LIVING
IN
SPACE

CULTURAL AND SOCIAL DYNAMICS, OPPORTUNITIES, AND CHALLENGES IN PERMANENT SPACE HABITATS

SHERRY BELL, PH.D.
AND
LANGDON MORRIS
EDITORS

FOREWORD BY
EDGAR MITCHELL, SC.D.

AN AEROSPACE TECHNOLOGY WORKING GROUP BOOK

THIS BOOK IS DEDICATED TO ALL WHO ARE COMMITTED TO
PEACEFUL ENDEAVORS IN OUR JOINT QUEST TO ACHIEVE
THE GOAL OF HUMANS LIVING IN SPACE!

• • •

THE EDITORS OF THIS VOLUME AND THE MEMBERS OF THE
AEROSPACE TECHNOLOGY WORKING GROUP WISH TO
THANK EACH OF THE AUTHORS WHO CONTRIBUTED THEIR
IDEAS TO THIS WORK.

OUR HOPE AND EXPECTATION IS THAT THESE BOOKS WILL
CONTRIBUTE TO OUR CONTINUING QUEST TO EXPLORE AND
DEVELOP OUR UNIVERSE AND OURSELVES, AND IN SO
DOING WILL DEEPEN OUR APPRECIATION AND
UNDERSTANDING OF THE OPPORTUNITIES THAT LIE
BEFORE US.

In the same series

Beyond Earth
The Future of Humans in Space
2006

TABLE OF CONTENTS

FICTION

CONCLUSION

THE AEROSPACE TECHNOLOGY WORKING GROUP

The Aerospace Technology Working Group, also known as ATWG, is an independent space policy research group led by seasoned professionals in aerospace and other fields who seek to further humanity's exploration of space while simultaneously benefiting people on Earth.

The ATWG was instituted by NASA Administrator Richard Truly in 1990 as an independent body to perform future planning for the nation's space efforts. Initially, the ATWG began identifying and seeking improvements in both existing and developing space systems through planned application of emerging technologies and the development of new ways of doing business, including the application of distributed missions and innovative operations strategic concepts.

Today, the ATWG is an independent entity, using semi-annual and regional Forums, technical and strategic dialogs, personal interactions, books, articles, and speeches to explore topics pertinent to developing a space-faring people and prepare policy recommendations for national and global leaders.

Using the organization's substantial base of management, engineering and scientific expertise, the ATWG also provides strategic and technical advice, public speakers, and consulting teams to address specific aerospace tasks and broad conceptual and philosophical questions.

Emphasis is on the use of systems engineering and system-of-systems engineering, while accounting for the broader effects on other industries, programs, the environment, and the day-to-day lives of this planet's inhabitants. The ATWG collaborates actively with other space-related national and international organizations.

In addition, the ATWG places special emphasis on promoting and stimulating education in the sciences, mathematics, the engineering disciplines, and other technical areas.

Participants in the ATWG forum meetings include experts from throughout NASA, aerospace contractors, systems suppliers, entrepreneurial businesses, professional societies, universities, and government agencies including the DOD, FAA, and DOE. Membership is open to professionals and professionals-to-be of any age, and as a reader of this book, you, too, are welcome to join.

You can learn more about the ATWG at www.atwg.org.

FOREWORD

EDGAR MITCHELL, SC.D.
APOLLO 14 ASTRONAUT

No words can adequately describe the experience of living in space, of living on another celestial body, of waking in the "morning" to the sight of Earth floating more than 200,000 miles away serenely surrounded by blackness, or skipping across a barren, dusty surface in 1/6 g. To date, only a handful of humans have had these life-changing experiences, and I am privileged to be among them.

When Alan Shepard and I walked - and briefly lived - on the moon in 1971, everyone involved in that endeavor expected that permanent habitats in space and on the moon would be common in at most a few short decades.

It has not turned out that way, but the promises and the challenges of space flight, space exploration, and space habitation have not faded from our sight. In fact we see today, even more than in the 1960s and 70s, that space activities hold important promise for humans on Earth as a source of needed energy, as a key element in the global economy, in the search for scientific solutions to Earth's environmental challenges, and even in protecting our beautiful ecosystem from space-borne threats.

For just as surely as the early Phoenicians began sea exploration in the Mediterranean thousands of years ago, and the South Sea Islanders in dugout canoes began exploring the vast Pacific, our generation has set the stage for humans to become explorers of the larger cosmos, perhaps in this century.

And in this regard, the book you are now reading makes a welcome and important contribution to the ongoing dialog about our destiny in space.

Living in Space is a captivating study across a wide spectrum of the issues that humanity faces, as we look beyond our home planet at future needs, future business endeavors, future learning opportunities, and future homes for our children and theirs. The authors who contributed to this volume present us with a wonderful diversity of perspectives, including the arts, philosophy, business, science, and technology, and the story that emerges from their fine writings engages the imagination.

These chapters also engage our vision, and I hope this book helps us to muster the will and the commitment to proceed with the development of space for the benefit of all humanity, as it should be. I hope you enjoy this book as much as I did.

Edgar Mitchell
March 26, 2009

INTRODUCTION

We are very privileged that such a distinguished group of contributors has come together in this volume to share their thoughts with us, and we thank each of them for taking the time to participate in this project.

In this book, you will find chapters covering a great breadth of themes, from broad philosophical, strategic, and conceptual overviews of the space endeavor, to focused technical discussions on topics ranging across the hard sciences, including space solar power, microbes, and inflatable habitats, as well as the social sciences, management, governance, and the arts.

This range demonstrates that our immanent journeys into space present challenges on many levels. The complexity that we face is clearly not just a matter of technology, but also a matter of culture, business, sociology, procreation, education, etc. In fact, we anticipate that all facets of human culture will eventually be adopted and adapted to conditions off the Earth, and thus the chapters in this book constitute a primer across a huge range of possibilities.

You will also find that this book includes both fiction and non-fiction chapters. We believe that the underlying and anticipated truths of fiction are important because they help us to envision human life in the entirely new set of circumstances that space presents, and that fiction therefore enables us to examine intriguing possibilities in new and useful ways.

It is our view, and the view of the ATWG as a whole, that one day humanity must, and inevitably will venture into space in significant numbers, and that when we do so, it must be for peace and human development rather than for war or domination. We hope that this book contributes to this broader intention, and that you are intrigued, challenged, and perhaps inspired by the chapters that follow.

EDITORS' ACKNOWLEDGMENTS

We wish to thank Rick Eckelkamp and Dr. Kenneth Cox of the Aerospace Technology Working Group (ATWG). They played an enormously important role in this undertaking. They were always available. Although they kept sight of the changes that were happening within ATWG, they continued to care about the people who were behind-the-scenes. In today's climate, this attitude was a breath of fresh air.

We also wish to thank the authors who contributed to this book. We learned a lot from them. Without them, there would be no book. Working with them has been an immensely gratifying experience, and we look forward to working with each of them again.

SHERRY BELL

I want to thank my good and kind husband, Dr. James Wakefield, my darling daughter Aubrey, and my charming grandson Laine for their patience and understanding throughout this enterprise. I also want to thank Langdon Morris. His expertise and previous experience with publishing proved invaluable.

I owe a debt of gratitude to Dr. Bob Krone. Dr. Krone has been my role model and rock of Gibralter throughout this project. Dr. Krone pioneered this series of books and is the Editor of the first volume. He has generously shared his advice and wisdom on many occasions. Furthermore, in the meantime, he has become a cherished friend.

Last but not least I want to thank my good friend Mark Hopkins of the National Space Society. Mark was a beacon of light during the dark times, and, as is usually the case in an undertaking like this, there were some.

I am honored to have had this opportunity. Thank you all!

LANGDON MORRIS

I want to thank my family for their support throughout this project, and I also want to acknowledge Sherry for her leadership in bringing this book to fruition.

NON-FICTION

CHAPTER 1

THE OVERVIEW EFFECT, THE COSMA HYPOTHESIS, AND LIVING IN SPACE

FRANK WHITE

WE ARE LIVING IN SPACE

In asking the question of whether people living in space will think or act differently from those living on Earth, we must first begin with a definition of what we mean by "living in space." The truth is that we are all living in space right now. The Earth is in space, it has always been in space, and it will always be in space. When we talk about "going into space," or "living in space," we are really talking about leaving the Earth and seeing the universe from a different point of view, a non-terrestrial one.

Those who leave the Earth and live in space habitats, on planetary surfaces, or in generational starships will not be different from those who remain on the Earth because they are living in space, but more likely because they will be far more aware of that fact!

Having clarified that very important issue, we are now free to ask ourselves what will be different about spacefaring civilizations as opposed to those that are Earthbound. In *Beyond Earth*, we explored a paradox: while the view of the Earth from orbit or the moon invokes a realization of our essential unity, exploration of the universe seems likely to create a diversity that might overwhelm a briefly realized sense of oneness. A positive way to see this development is to call spacefaring society "pluralistic," which means that it will contain a rich variety of perspectives. On the other hand, as with similar societies on Earth, we need to ask ourselves what will hold this new civilization together? At what point does pluralism degenerate into conflict, or even civil war?

THE OVERVIEW EFFECT

My own understanding of the Overview Effect and its potential impact on those living permanently in space emerged on a cross-country flight some three decades ago. I was thinking about the changes that might take place among those who would live in O'Neill-like space habitats, and as I gazed out the airplane window at the Earth below, I thought, "People living in space settlements will always have an overview. They will know intuitively what we have been trying to understand intellectually for centuries: all life on Earth is interconnected, and we are, in a very real sense, all one." This insight remains true, and seems to be confirmed when we talk with those who have lived, however briefly, in Earth orbit or on the moon.

Nevertheless, it seems highly unlikely that human beings will be content to remain in our near-Earth neighborhood. Members of the space movement talk with enthusiasm about free-floating space habitats at Lagrange points, moon bases, terraforming Mars and Venus, and launching generational starships to leave the solar system and head for the stars (no matter how long it takes to get there).

Even as we explore just the solar system, we can imagine dramatically different cultures, philosophies, and political systems arising out of human migration away from the home planet. Settlements on Mars might, for example, break away from the terrestrial governments that established them. For these people, seeing the Earth from 35 million miles away may yield a different reaction than seeing it from orbit or the moon. They may feel, as American settlers felt about the King of England, that those remaining on the Earth could not possibly understand the conditions they faced on Mars, and this would naturally lead them to declare political independence.

Or someone meditating on the moon might decide that he or she has had a realization that lays the foundation for announcing a new religion that would come into conflict with the religions of Earth. (I met some people at one space development conference who wanted to make space exploration itself into a religion. I have lost touch with them, and I do not know if they have continued their quest, but this is an example of the new thinking that space exploration naturally engenders.)

Moreover, we do not know the long-term impact that living in a space habitat's weightless environment would have on the human brain, yet this experience might well result in subtle changes in the thought processes that we cannot now envision. Those small changes might cascade into deeper transformations that we will never be able to predict based on the Earth-based science of today.[1]

THE RISE OF HOMO SPACIENS

As I noted in my contribution to *Beyond Earth*, the most dramatic shift that we can imagine might well involve speciation, i.e., a natural evolution, in response to radiation, varying gravity environments, and isolation from the main gene pool, of homo sapiens into homo spaciens. Some have argued that the likely outcome of space settlement is not the evolution of one, but rather of many, such new species.[2]

Even today, without widespread space exploration under way, there is a movement afoot among Earthbound humans to use technology to propel our species to its next level of evolution and into a "posthuman" era. We can imagine that as human beings living in various space habitats make the fundamental choice never to return to Earth, they may choose (or be forced) to adapt themselves more fully to the environments in which they find themselves.

In considering all of these possibilities, we must remind ourselves that what we call the "space environment" is not a single homogeneous thing, and that people will also need to make decisions as to whether they want to adapt to the new environment or rather adapt the new environment to themselves.

Let's consider, for example, settlers on Mars who plan to rarely, if ever, return to Earth. They find themselves on a planet similar to Earth in many ways, such as its 24-hour rotation period, a landscape that looks a lot like Arizona, and temperatures that are extreme, but tolerable compared

---

[1] This issue has been explored in some detail by George Robinson, fellow ATWG member and *Beyond Earth* author.

[2] As I have noted before, many of my ideas on speciation come from an interview with Peter Diamandis in the summer of 1986. Ben Finney and Eric Jones also gave a seminal talk at the Space Studies Institute in 1983, which I quoted in *The Overview Effect*, p. 157. Their complete speech appears as "From Africa to the Stars: The Evolution of the Exploring Animal," Space Manufacturing 1983, *Advances in the Astronautical Sciences, Proceedings of the Sixth Princeton/SSI Conference on Space Manufacturing*, edited by J. D. Burke and A. S. Whitt, Vol. 53, Univelt, San Diego, pp. 85-100.

with other planets in the solar system. Gravity is less than that on Earth, and perhaps little can be done about that, but our new Martians might be able to take many steps short of terraforming to create a lifestyle on Mars that is not very different from that which they lived on Earth. Except for the previously discussed likelihood of breaking away from terrestrial government, they might even reproduce many of the political systems they brought with them.

Terraforming certainly could take the process much further, creating water in abundance, an atmosphere that humans could breathe, and seasons that mimicked what the settlers had experienced back home.

Another possibility open to the new Martians, on the other hand, is 'xenoforming,' adapting themselves to the Martian environment. Like the inhabitants of the fictional planet of Dune, they might learn to live with much less water, and through genetic engineering, or the use of other technologies, find a way to live on Mars as it is, rather than adapting it to Earthly preferences.

These scenarios are explored brilliantly in the trilogy, *Red Mars, Blue Mars, Green Mars*, by Kim Stanley Robinson. In his thoughtfully constructed examination of how human society might evolve on the Red Planet, he pictures both options being played out by different factions of settlers. Some devote their energies to massive terraforming projects, while others melt away into the Martian "outback" and become "true Martians." As one might expect, the two groups find little in common, and conflict inevitably arises.[3]

This scenario seems plausible in part because we can see it happening on Earth today. Some human cultures would prefer to live in harmony with the planet as it is, eschewing heavy use of technology, and trying to adapt to the environment rather than vice versa. Others tend to see the environment as something to be shaped to human ends. Recently, with rising concerns about climate change and global warming, those in the environment-shaping cultures have begun to look more closely at ways to live in harmony with the Earth system. This growing environmental awareness is, I believe, at least partially a byproduct of space exploration and the Overview Effect. Having seen the Earth from space, we now know it is a finite, interconnected system, and that we are an integral part of it. Everything we do on the Earth must be understood in terms of its impact on a delicately balanced system.

A central question posed in the Mars trilogy is whether living in space will lead to human beings becoming fundamentally better in some way. Will we leave behind the divisions and considerations that separate us and find ways to live together in greater harmony? While there is no

---

[3] Robinson, Kim Stanley, *Red Mars*, Spectra, 1993.

definitive answer, the trilogy clearly shows the challenges to this notion through its characters and their interactions.

THE NEW OVERVIEW EFFECT

As thousands of people prepare to go into space as customers of private space companies, we must now pose the question in practical terms, not as part of a novel: if the Overview Effect experience, derived from looking back at the Earth from orbit or the moon, reminds us that the barriers between us are a matter of perception; what happens when our gaze turns the other way, out into the vast universe that lies before us and is ready for our exploration and understanding of it? (As my colleague David Beaver points out, seeing the Earth in space may be as important as seeing it from space. Moreover, turning away from the planet and gazing into the cosmos may be even more profound.)

In my book, *The Overview Effect*, I suggested that the act of exploration itself might become the unifying force that would hold us together through light-years and millennia. I proposed that we create a central project, like the building of the great Gothic cathedrals, to last for a thousand years, and that this central project be known as "the Human Space Program." This monumental effort to understand the universe and the human role in it would be enough, I thought, to engage the energies of millions of people. It would, if understood properly, be a unifying force as we evolve into the universe.

In writing a chapter for *Beyond Earth*, I concluded that this idea, while still valid, needed to be modified to take into account the many scenarios that might unfold as a result of speciation, genetic engineering, evolution of societies on other worlds, and even contact with extraterrestrial life and intelligence. Accordingly, I proposed that the Human Space Program become a Post-Human Space Program that would be broad enough to include any form of intelligence that we might meet, or become, over time.

In writing this chapter, I have concluded that we need to consider one further expansion of the idea, which is to begin thinking about the nature of the universe itself. Of course, the point of exploring the universe is to understand it through experience, not to prejudge what we will find. Taking the latter approach might lead us to make the same mistake Columbus made when he "discovered" a land that he thought was India, which later became known as "America." It had long been inhabited and discovered by others who are often called "Indians" to this day because of his geographical error.

THE POST-HUMAN SPACE PROGRAM

Nevertheless, I do think it is worthwhile to put forward a working hypothesis about the environment we will be exploring with the Human or Post-Human Space Program.   And I believe a hint of that hypothesis is found, yet again, in the reports of our astronauts and cosmonauts. Even as they gazed back at the Earth while in space, they also looked past our home planet and saw the vast universe awaiting us.

Some of them experienced what I call "the Universal Insight." Edgar Mitchell, for example, has said:

> ...it gets you closer to a more universal experience because of the distance and wider view. You identify more with the universe as it is instead of the Earth as it is.[4]

One of the early shifts in consciousness resulting from the overview experience came to us in the form of the "Gaia Hypothesis."   As formulated by James Lovelock, the hypothesis suggests that the Earth is not simply a dead planet with living creatures riding through the universe on its surface. Rather, Lovelock saw the planet as a living system in which life plays a huge and critical role. He speaks of Gaia (Earth) as almost a sentient being.[5]

THE COSMA HYPOTHESIS

In the process of understanding the Overview Effect and Universal Insight more completely, the term "Cosma Hypothesis" came naturally to mind for me. It would seem that what we have said of Gaia — the Earth — we could as easily say of Cosma, the universe.

Even if life and intelligence have so far been found on only one planet, the cosmos can be said to be, to some extent, alive and conscious. Our expectation is that there may be thousands, or even millions, of planets like our own that have given birth to entities like ourselves. If so, the great adventure of our lives will be linking up with others like ourselves. And if not, one of the purposes of human space exploration may well be to transmit life and intelligence to regions where these do not yet exist.

Here, then, lies a new form of unity. The universe, or Cosma, is, like the Earth, an interconnected and interdependent system. We normally see ourselves as a tiny, perhaps even insignificant part of that system, but we

---

[4]   White, Frank, *The Overview Effect*, AIAA, Reston, VA, 1998, p. 203.
[5]   Lovelock, James E. *Gaia: A New Look at Life on Earth*. Oxford: Oxford University Press. 1979.

do not know for certain that this is true, and the only way to find out is through what we call "space exploration," and what might better be called "evolution into the universe."

The idea that the cosmos is, to some extent, alive and conscious, is not new. Indigenous peoples, for example, find our efforts to distinguish between conscious and non-conscious and living and non-living things somewhat limiting.[6]

The great Oxford/Cambridge don and Christian thinker, C. S. Lewis, imbued his science fiction with the same kind of thinking. In *Out of the Silent Planet*, his protagonist, Ransom, finds "space" to be something altogether different than he had imagined:

> He could not call it "dead;" he felt life pouring into him from it every moment. How indeed could it be otherwise, since out of this ocean the worlds and all their life had come?[7]

In more recent times, we find our own contemporary thinkers like Barbara Marx Hubbard, Steven Wolfe, and Duane Elgin voicing similar ideas. Over time, then, it would seem that space exploration is becoming more important in terms of how we think and how we experience the journey, and less important in terms of what means of propulsion we use or how we get there.

Many will of course wonder how all of this relates to our understanding of God, or the Divine, and there are likely to be as many answers as there are questioners. I have come to the conclusion that, for me, space exploration means many things, but at the spiritual level, it represents an attempt to understand the Creator by understanding the creation. I know that this is an imperfect means to comprehend that which is beyond comprehension, but it seems worth the effort, nevertheless, and one that I will gladly undertake while living in space, just as I have while living on Earth.

It seems likely to me that each human being who chooses to live in space will one day ask similarly deep questions about the nature of our existence, just as we have been inspired to do on Earth. Will they verify or deny the Cosma Hypothesis? Let's find out.[8]

---

6   White, Frank, *The SETI Factor*, Walker and Company, New York, 1990.
7   C. S. Lewis, *Out of the Silent Planet*, Scribner Paperback Fiction, New York, 1996, p. 32.
8   A group of colleagues and I are now undertaking to create the Overview Institute to explore these issues by studying the Overview Effect as well as disseminating it worldwide. We hope to open a dialogue about issues surrounding living in space, human purpose, and the evolution of consciousness.

FRANK WHITE

Frank White is the author of *The Overview Effect: Space Exploration and Human Evolution*, first published in 1987 and re-issued in 1998. Based largely on interviews with astronauts, *The Overview Effect* documents the space flight experience, especially the impact on consciousness of seeing the Earth from orbit or the moon.

A member of the Harvard College Class of 1966, Frank concentrated in Social Studies, graduated magna cum laude, and was elected to Phi Beta Kappa. He attended Oxford University on a Rhodes Scholarship, earning an M.Phil. in 1969. He is the author or co-author of five additional books on space exploration and the future, including The SETI Factor, Decision: Earth, Think About Space and March of the Millennia (both with Isaac Asimov), and The Ice Chronicles (with Paul Mayewski). He also contributed chapters on the Overview Effect to two additional books on space exploration, Return to the Moon and Beyond Earth.

Frank has also spoken at numerous space-related conferences. In 1988, he delivered the keynote address at the International Space Development Conference in Denver. In 1989, he spoke at George Washington University to mark the 20th anniversary of the Apollo 11 moon landing. In 2006, the Space Tourism Society awarded Frank a "Certificate of Special Recognition." In 2007, he delivered the keynote address at the first Overview Effect conference in Washington, D.C. Frank is now working with several associates to create the Overview Institute, which will conduct continuing research on the Overview Effect and share the findings as widely as possible.

He is married to Donna White. They have five children and three grandchildren.

CHAPTER 2

EARTHLINGS ON MARS:
THE PHYSIOLOGICAL PSYCHOLOGY OF CULTURAL CHANGE

DAWN L. STRONGIN, PH.D.
CALIFORNIA STATE UNIVERSITY, STANISLAUS
AND
E. K. REESE
CALIFORNIA STATE UNIVERSITY, STANISLAUS

A society is defined by its boundaries, a culture by its horizon.
James Carse

Much of human culture serves two purposes: it provides the means to meet the physiological requirements for sustaining health and life, while simultaneously satisfying our psychological needs for meaning and purpose. Culture has a strong influence on our behaviors, which helps us to find our feet in new and difficult circumstances. In fact, extreme environments challenge our species to invent new ways to thrive, for in the words of Nietzsche, "He who has a why to live for can bear almost any how."

Technology now makes it possible for humans to journey farther in three dimensions than ever before, satisfying a wanderlust beyond the pull of gravity and toward life in space and in other worlds. The international community's intellectual, financial, and creative resources are currently focused on Mars as the first space settlement. If this vision is realized, Mars may be the next world to host human culture. However, we should not assume that Martian culture will be identical to what humans have developed on Earth. What the Martian culture will become is dependent upon what the pioneers contribute to their new world in the face of novel and daunting challenges.

Pioneering precedence suggests that these settlers' cultural transformation will be based on their prior experiences and the behavioral adjustments they make to cope in their new physical environment. Over generations, then, the Martian dweller may become quite a different type of human.

THE PIONEERS

The pioneers may be a diverse group, but they are sure to share some common traits. Studies of those who choose to venture into extreme conditions tend to fall into one of three categories (Johnson & Holbrow, 1975). The first includes those whose romantic ideal of rugged exploration eventually exceeds their realistic capacity to endure intense discomfort and/or isolation. The insurmountable stress of living in these conditions leads to physical and psychological suffering, and these individuals turn their dreams toward the first ride home.

The second category of pioneers includes individuals who seek only fortune. They are more likely to tolerate difficult conditions to achieve their goal, but as they are not committed to the environment as a part of their home, their opportunism may preclude them from helping to develop a fulfilling culture.

The last category of pioneers is what physiological psychologists identify as Sensation Seekers. They find new challenges highly rewarding, and thus they actively pursue novel, intense, and complex sensations and experiences. Adventure Seekers are a subcategory of Sensation Seekers who hunt for stimulation through exploration. They are expected to be the most successful space pioneers because they are most likely to thrive under the difficult conditions of prolonged space travel and settlement. In addition, they will likely find the challenges involved in establishing a unique society and culture gratifying.

However, Adventure Seekers may also be more vulnerable to boredom. The predicted six-to-eight-month flight to Mars will necessarily involve confinement in tight quarters and unrelenting monotony. This condition would be particularly distressing for Adventure Seekers, a group who require stimulation above and beyond what is necessary for the average person.

American polar explorer Admiral Richard Evelyn Byrd addressed the psychological effects of prolonged confinement in his 1938 book, *Alone*:

> *There is no escape anywhere. You are hemmed in on every side by your own inadequacies and the crowding measures of your associates. The ones who survive with a measure of happiness are those who can live profoundly off their intellectual resources, as hibernating animals live off their fat.*

Recognizing this situation, we could imagine that spacecraft designers might include features such as immersion virtual reality simulations to keep the voyagers occupied during the long months of their journey.

HETEROGENEITY

Colonists on Mars will work together extensively on both the building and maintenance of all systems on their colonies, and thus they will need to work well as a group. Marilyn Dudley-Rowley's research on social interactions in Arctic, Antarctic, and space expeditions found that larger, more culturally heterogeneous groups who worked together for longer periods of time performed better in extreme conditions when compared to smaller, homogenous groups who worked together for shorter time periods (Dudley-Rowley, 2004). These findings should be considered during the selection of colonists, as they suggest that broad cultural diversity may help mitigate social tensions.

The cultures represented among the first pioneers on Mars will have the largest impact on how culture evolves there, while Earthbound nations who join the Mars colonies later will likely influence the growing culture to a lesser degree. A large, diverse population on Mars to begin with will give the colonies a much broader cultural palette with which to create and shape a vibrant culture in this new frontier.

MICROGRAVITY

One of the essential challenges the space traveler faces is weightlessness (in space) or low gravity (.33gEarth on Mars). The development of human physiology on Earth has relied on Earth's gravitational effects. Gravity shapes life and culture; therefore we can anticipate significant differences on Mars.

While human physiology defines the character of our sensory system, nervous system, reflexes, and bones (Shwartz, 2002), it nevertheless adapts relatively quickly to the absence of gravitational resistance. For example, within two weeks of weightlessness, spacefarers recover from Space Adaptation Syndrome (SAS), a condition much like sea sickness, in which incongruent information from the eyes, vestibular system, and gravity sensors induces nausea and general feelings of malaise.

A series of experiments on the Space Life Sciences Mission flown in 1991 was designed to identify what the specific changes in the nervous system are that allow spacefarers to adjust. Ross (1996) found that neurons (i.e., brain cells) in rats developed much denser and more elaborate connections with each other at approximately the same rate they adjusted to weightlessness. This finding suggests that these new connections (i.e., synaptogenesis) are the nervous system's method of adapting to different conditions.

As social creatures, humans connect with each other through shared sensory experiences, which in turn serve to facilitate survival and a sense of well-being. As sensory experiences change, so do the methods of social connection. For example, compared to all other sensory modalities, taste is the most social sense in our current cultural context, often experienced in a setting shared with others. Food is an important component in every culture, with many daily activities and interactions revolving around it. Most social traditions involving shared emotions also involve sharing food – from commemorations, family mealtimes, and simple gatherings to worship and grieve. It serves to comfort, soothe, and please.

Our intricate ability to taste is due, in part, to gravity. We have receptors for only five tastes: sweet, sour, salt, bitter, and umami (sometimes translated as savory). All other "flavors" are actually aromas. However, only volatile substances emit a scent, and as molecules evaporate we sniff to move them closer to the moist, fatty olfactory bulb in the nasal cavity behind the bridge of the nose. Mucosa containing receptor cells absorb these molecules.

Humans have approximately five million olfactory nerve endings in the nasal cavity, yet a response from only 40 nerve endings permits us to consciously process a scent. To savor food more fully, we exhale with slightly opened mouths, moving the molecules to the olfactory bulb to further accentuate its taste and smell (Ackerman, 1990).

All of this is experienced differently in space. The importance of gathering during mealtimes has already been incorporated into space culture. Today's spacefarers use footholds to remain stationary around a central table for meals, but one of the biggest complaints of spacefarers is the loss of flavorful pleasures. Reduced gravitational force dampens molecular volatility, and thus food loses its aroma. Because the five tastes do not require molecular volatility, spacefarers have heightened sensitivity to the basic tastes only, leaving something to be desired in the formerly complex flavors of their favorite cuisine. The deficits in the sense of smell result in one consistent complaint among spacefarers: once appealing foods become tasteless at best, and like putrefaction at worst. (Imagine attempting to enjoy a meal with a perpetual head cold!)

Spacefarers come to crave foods saturated in piquant flavors, regardless of their preference for spicy foods on Earth. Consider the effect of capsaicin, the active ingredient in chili peppers. It has no flavor or aroma, but it triggers a burning sensation, and endorphins (endogenous morphine) are released to ameliorate the pain. The endorphins also promote euphoria (or to a lesser extent, mild pleasure). Spacefarers find that the experience of endorphins drenching the fire partially compensates for the lost pleasure of complex flavors.

In this setting, prolonged reduction of smell will eventually cause the olfactory system to atrophy. We can anticipate that the nervous system will attempt to compensate by establishing more elaborate neuronal connections in active areas of the brain, thus creating greater sensitivity in other sensory modalities. Perhaps this sensory encroachment will change within the species as it adapts over evolutionary time.

Yoav Gilad and his colleagues at the Max Planck Institute for Evolutionary Anthropology in Germany and the Weizmann Institute in Israel found a relationship between the loss of olfactory receptor genes and the acquisition of full trichromatic color vision in human evolution. Today, a high percentage of the human olfactory DNA sequence is no longer active. Following the "use-it-or-lose-it" rule, many of these olfactory receptor genes apparently became inactive as we pulled our noses up from the ground and stood on our hind legs.

In parallel, humans developed increased dependence on other senses, such as sight and hearing. Gilad and colleagues (2004) propose that the evolution of color vision coincides with diminished sensitivity to smell, suggesting that the spacefarers' diminished smell sensitivity may lead to further sensitivity gains in other forms of perception.

Pleasure formerly found in taste may be replaced by heightened visceral, visual, and tactile experiences. For example, the Adventure Seeking pioneer may find excitement in the Martian equivalent of fugu, or pufferfish. The pufferfish contains tetrodotoxin, one of the deadliest neurotoxins on Earth. The smallest taste creates a pleasurable tingling sensation on the lips and tongue, sometimes referred to as the "taste of death." However, just a little too much induces paralysis in all muscles, including the diaphragm, but, because the toxin never reaches the brain, the gourmand is fully conscious while enjoying this sensory adventure, with the possibility of lethal suffocation heightening the excitement.

Conversely, the decreased smell sensitivity in spacefarers and settlers may be advantageous to their health. Researchers have found that calorie restriction reduces insulin levels and slows metabolism – changes that are thought to slow the effects of age-related cell damage and increase longevity (Heilbronn & Ravussin, 2003). Reduced sense of smell may therefore be an effective appetite suppressant, with the desirable side effect of prolonging life.

However, given the harsh conditions inherent in space travel and likely in Martian settlements, it will be essential that spacefarers are adequately nourished to cope with the challenges of maintaining their health. Fabrizio, Pletcher, Minois, and Vaupel (2004) discovered that the lifespan of flies increased simply by uncoupling smell from taste, even without limiting calorie intake. It appears that the sensory system detects the availability of nutrition and regulates hormonal output based on this sensory input.

BONE DETERIORATION

Weightless and low gravity conditions also affect our bones. Although rigorous exercise in space will prevent the excessive loss of muscle mass, sustaining bone density requires additional gravitational resistance. Like most cells in the human body, bone cells go through a cycle of regeneration, resorption (or degradation), and formation of new cells, and bone mass regeneration is activated in response to weight-bearing signals. Hormones and growth factors stimulate osteoblasts to produce the collagen that forms a new bone matrix; but in weightless conditions, resorption occurs faster than the formation of a new bone matrix, resulting in bone loss.

A four-year study found that spacefarers on the International Space Station lost as much lower body bone mass in one month as an elderly woman loses in an entire year (Roy, 2006). Traveling to Mars for six months in this weightless condition would presumably reduce bone mass by one-third. Although bone mass can be replaced within one year after return to Earth's gravity, the bone structure and density in the weight-bearing bones cannot recover, leaving the bones more massive and yet more prone to fracture. However, bone structure in the non-weight bearing upper body remains intact during prolonged weightless conditions (Roy, 2006).

Bone density will be further compromised by radiation exposure in space and on Mars. Earth's magnetosphere (which Mars lacks) shields it from the torrential million miles per hour solar winds. As spacefarers leave Earth's protection, they become vulnerable to the bombardment of bone-stripping, cancer-inducing radiation. This combination of factors will most likely evoke new and unforeseen adaptive behaviors.

GENETIC ENGINEERING

One solution to the problem of high radiation levels and other inhibiting conditions on Mars might include genetic engineering. The polyextremophilic bacterium Deinococcus radiodurans might hold the key. D. radiodurans, which is the most radioresistant organism known to exist, is also able to withstand conditions of extreme high and low temperatures, dryness, vacuum, and acidity. Further study of the structure and composition of this organism's genetic material may provide insight into how human beings and other organisms might be genetically altered for better suitability to the Martian environment. Especially pertinent is the possibility that this organism may, in fact, have originated on Mars and traveled to Earth on a meteorite, as some scientists suppose (Pavlov, Kalinin, Konstantinov, Shelegedin, & Pavlov, 2006).

The subject of genetically engineering terrestrial humans into Martians brings along with it the question of how these altered beings might be viewed by the rest of society. Would they be admired as a more advanced life form, or be avoided, shunned, and feared as a type of "Frankenstein's monster?" Would they be treated as citizens, with choices and inalienable rights, or be viewed as science projects, owned and ruled by their "creators?" With such drastic genetic transformations, would they even be considered human? Their possible classification as "non-human" could be used in defense of unchecked experimentation, abuse, and cruelty, and the resulting isolation would be another factor distinguishing Martian culture from Earth's. An abundance of science fiction has explored this topic; perhaps now we will find out whether it is fact or fiction that is stranger.

SOLIPSISM

Radiation bombardment on Mars will necessitate well-protected, enclosed habitation, thus eliminating natural vistas and light. This poses the risk that Martian inhabitants will experience solipsism syndrome, a state of mind characterized by the sense that everything surrounding an individual, the entire environment, is not real, and that all existence is a long dream from which one cannot awaken (Johnson & Holbrow, 1975).

Solipsism syndrome seems to stem from situations in which there is too much predictability in the environment, where everything in a person's surroundings is "unnatural," controlled, and unchanging. The mindset that reality itself is unreal can take over, and the affected person becomes apathetic and indifferent.

Another potential psychological complication for those living away from natural light is Seasonal Affective Disorder (SAD). This disorder is prevalent in areas of persistent darkness, such as in the long Alaskan, Scandinavian, and polar winters. Under non-polar conditions, the rhythm of light and darkness stimulates the production of melatonin to promote sleep, but the natural circadian rhythm is interrupted by chronic darkness, creating an overproduction of melatonin, and leaving individuals prone to symptoms of severe depression and anxiety. These symptoms can best be addressed by light therapy.

Daily exposure to an hour of high intensity full spectrum lights helps to maintain circadian rhythms (Gagné, Gagné, & Hébert, in press). Lighting can also help mitigate feelings of solipsism if they simulate a natural and uncontrolled environment, such as artificially created "sunlight", which can be used to facilitate the illusion of an outdoor environment. This illusion could be further enhanced by projections of the natural lighting and outside weather patterns into a settlement's ambience.

These issues pertaining to day-to-day "reality" must be addressed in the architecture of any extraterrestrial colony or space-bound structure. They must provide a realistic, seemingly uncontrolled "natural" habitat, lest our minds figure out this farce.

THE PASSAGE OF TIME

Differences between the Earthly and Martian calendars could have a large impact on the settlement's emerging culture. The Martian day is about 24 hours and 37 minutes, a little longer than Earth's 23 hours and 56 minutes. Mars' full orbit around our sun is equivalent to approximately 687 Earth days. Mars' elliptical orbit makes Martian years and seasons nearly twice as long as their Earthly counterparts.

These differences, though seemingly inconsequential at first glance, have the potential to alter culture in drastic ways. For example, how would holidays be celebrated? If Martians were to decide to rely upon the Earthly calendar, they would celebrate Chanukah/Christmas/Kwanzaa/Solstice (in alphabetical order) during winter in one year and summer in the next. It is reasonable to assume that, over the passage of time, new holidays would develop and older (Earthly) ones would lose significance.

The concept of age would also be affected. A person living on Mars lives one Martian year in the same period as his sibling on Earth has lived two. Thus, a 50-year-old is only 25. Years would be translated like languages in the future of interplanetary travel.

However, it is currently unknown how the combination of low gravity, high radiation, calorie restriction, and other foreseeable and unforeseeable aspects of Martian life may affect the human lifespan. A significantly shorter or longer life expectancy would therefore alter our views and assumptions concerning age and ageing, rendering the translation of Earth years into Martian years moot.

SEXUALITY AND REPRODUCTION

Weight-bearing bone loss and muscle atrophy of the extended low-gravity condition may make sexual intercourse and human childbearing, as we currently know them, impossible. When considering brittle bones, creative Martian sexual practices would need to replace "dangerous" Earthly coitus. Certainly, the Sensation Seeking settlers would put imagination and technology to good use.

Unless pregnant women lived in zero gravity chambers or large bodies of water, the compromised pelvic bone integrity would make carrying a child to term difficult, even with Mars' lower gravity. Natural childbirth would be prohibitive. Test-tube babies might therefore become a necessity, creating completely novel cultural associations concerning parenthood. The ramifications are far-reaching, including changes in gender roles, as women would no longer be the bearers of children, and unplanned pregnancies would be nonexistent.

Unfortunately, child bearing could become more expensive, giving rise to a social hierarchy in which only the wealthy (or those deemed suitable by the hierarchy's pinnacle citizens) could procreate.

INTO THE UNKNOWN

Beginning with our first Martian colony, some challenges will be predictable and easily controlled for; others will present unexpected difficulties. Innumerable obstacles and complications will certainly arise in our quest to explore the far reaches of our universe. Careful consideration of the physiological and psychological intersection that gives rise to culture will enrich blossoming colonies on Mars, and will serve to enhance the likelihood of success in this unprecedented frontier pursued by 'unreasonable' men, women, and children.

The reasonable man adapts himself to the world;
The unreasonable man persists in trying to adapt the world
to himself.
Therefore, all progress depends on the unreasonable man.
                                        ~ George Bernard Shaw

DAWN L. STRONGIN, PH.D.

Dawn L. Strongin, Ph.D., is an Associate Professor of CSU Stanislaus, specializing in cognitive neuroscience. She earned her Ph.D. at the University of Northern Colorado in Neuropsychology, Applied Statistics, and Research Methods, and minored in Educational Psychology. Dr. Strongin's areas of research include neurotoxic effects of organophosphates and solvents, and psychometric development. When not teaching, researching, and traveling, she is dreaming...of life on distant rocks.

E. K. REESE

E. K. Reese is a graduate student at Miami University working on her Ph.D. in clinical psychology. She earned her B.A. in psychology from California State University, Stanislaus. Ms. Reese's research interests include mental illness stigma and constructivist psychotherapy processes.

REFERENCES

1. Ackerman, D. (1990). *A Natural History of the Senses*. New York: Vintage Books/Random House, Inc.
2. Byrd, R. E. (2003). *Alone*. Washington, DC: Island Press/Shearwater Books. (Original work published 1938).
3. Dudley-Rowley, M. (2004). "The Great Divide: Sociology and Aerospace." Paper presented at the meeting of the California Sociological Association, Riverside, CA.
4. Fabrizio, P., Pletcher, S. D., Minois, N., & Vaupel, J. W. (2004). "Chronological aging-independent replicative life span regulation by Msn2/Msn4 and Sod2 in Saccharomyces cerevisiae." FEBS Letters 557, 136-142.
5. Gagné, A., Gagné, P., & Hébert, M. (in press). "Impact of light therapy on rod and cone functions in healthy subjects." Psychiatry Research.
6. Gilad, Y. (2004). "Evolution of primate sense of smell and full trichromatic color vision." PLoS Biol 2, 1: e33 doi:10.1371/journal.pbio.0020033
7. Hamilton, S. A., Pecaut, M. J., Gridley, D. S., Travis, N. D., Bandstra, E. R., Willey, J. S., Nelson, G. A., & Bateman, T. A. (2006). "A murine model for bone loss from therapeutic and space-relevant sources of radiation." Journal of Applied Physiology, 101, 789-793.
8. Heilbronn, L. K., & Ravussin, E. (2003). "Caloric restriction and aging: Review of the literature and implications for studies in humans." American Journal of Clinical Nutrition, 78, 3, 361-369.
9. Johnson, R. D., & Holbrow, C. (Eds.). (1975). Space settlements: A design study. NASA SP413.
10. Pavlov, A. K., Kalinin, V. L., Konstantinov, A. N., Shelegedin, V. N., & Pavlov, A. A. (2006). "Was Earth ever infected by Martian biota? Clues from radioresistant bacteria." Astrobiology, 6, 911-918.
11. Ross, M. D. (1996). "Gravity sensor plasticity in the space environment." Paper presented at the NASA Ames Research Center Astrobiology Workshop.
12. Roy, S. (2007). "How long does it take to rebuild bone lost during space flight?" Retrieved March 26, 2007, from http://www.nasa.gov/mission_pages/station/science/subregional_bone.html
13. Shwartz, M. (2002). "Human biology professor seeks participants for NASA hypergravity experiment. Stanford report." Retrieved March 25, 2007, from http://news-service stanford.edu/news/2002/march20/centrifuge-320.html

CHAPTER 3

OVERVIEW GOVERNANCE

BOB KRONE, PH.D.

INTRODUCTION

Space benefits for Earth and its humanity will create solutions to problems on Earth. Many of the impacts of problems are increasing with time and new ones have emerged in the 21st Century. Space commerce is predicted to become the biggest business in history. The ultimate solution for Earth's growing energy needs is space-based solar power. Once harnessed, it will be both a never-ending power source, and also dramatically reduce atmospheric carbon pollution. Energy and economy problems have forever caused conflict and trauma on Earth. There is building support for the theory that even peace on earth can result from the human migration to space.

But an optimistic future is not assured. In fact, the failure to solve the problem of Earth's national and international governance for the exploration and exploitation of space, and for human settlements in space will guarantee a future that benefits no one. This chapter provides a recommended governance model following the 2006 analysis by Yehezkel Dror.[1]

---

[1] Yehezkel Dror, "Governance for a Human Future in Space," Chapter 5 in Bob Krone, Ph.D., Editor, *Beyond Earth: The Future of Humans in Space* (2006, Apogee Space Press).

HOW TO GOVERN HUMAN FUTURES IN SPACE

Frank White has given the world a profound explanation for why space is different from Earth; he calls it the *Overview Effect*.[2] He points out that when humans saw the Earth for the first time from off the planet, they saw a small blue ball surrounded by the limitless expanse of space. From that unique perspective, the definition of our planet was immediately and permanently changed. Suddenly a deeper unity was revealed: "...all life on Earth is interconnected, and we are, in a very real sense, all one."

Frank also reminded us of a fundamental question that is examined in the Mars Trilogy[3] novels and at his own Overview Institute, one that we all must contemplate: *"Will living in space lead human beings to become fundamentally better in some way?"*

The forty-two co-authors who contributed to the previous volume in this series, *Beyond Earth: The Future of Human in Space*, as well as the many authors who have contributed to this volume, all agree that the answer to that question is, 'Yes!'

But we are not utopian thinkers. In fact, we may be among the better qualified to understand the serious hurdles that humanity must overcome before we obtain the social, biological, political, commercial, and economic benefits for Earth that may come from our endeavors in space.

And one particularly critical system required to bring the massive resources that await us in space to Earth, is effective governance. As Yehezkel Dror points out, "Without governance direction, support and supervision, there will be no human settlement in space."[4]

OVERVIEW SPACE GOVERNANCE MODEL

The many questions concerning the need for human space exploration and habitation have been debated for decades; the short answer to many of them is our view that human survival will eventually depend on building civilizations in space.

Our planet's resources are limited, and there are natural and human-made threats to Earth-bound life that are increasing in both probability and magnitude over time. The need for increased quality of life for billions of humans, and at the same time the need to prevent the erosion of today's

---

[2] See Frank White's chapter in this book, *"The Overview Effect, the Cosma Hypothesis, and Living in Space."* His original publication was *The Overview Effect*, (AIAA, Reston, VA, 1998).

[3] Kim Stanley Robinson (1993) *Red Mars*, Spectra.

[4] Yehezkel Dror, *"Governance for a Human Future in Space."* Chapter 4 in Bob Krone, Ph.D., Editor *Beyond Earth: The Future of Humans in Space*, Apogee Space Press. 2006, p. 42.

qualify of life, demands that we learn to capture resources known to be in space.

Science and technology will advance to make human settlements in space feasible within the 21st Century. So in answer to the question, Why Go? we already have. But answers to the questions How? and When? are still being formulated. To this end, space scholars and practitioners conclude that the most critical accomplishment for humans to attain in order to successfully create settlements in space...is effective governance.

In other words, human settlements in space will begin a completely new epoch for humanity, but how will they be governed?

Ross W. Ashby's Law of Requisite Variety[5] states that to comprehend and respond to a message requires that the recipient can apply at least the same scope of variety as the message itself contains. As this law applies to human endeavors on Earth, so will it also apply to the governance system for humans in space. Hence, political, social, cultural, environmental, management, economic or health problems that arise within communities in space will be unsolvable, unless the governance systems used in space are competent to match the massive complexity of the problems. This is not an insignificant challenge.

This chapter takes a large leap across the huge body of literature, including science fiction, to present some initial considerations for the design of a normative, generic governance system for human civilization in space. Drawing again on Frank White's work, I propose for it the name, "Overview Space Governance."

ASSUMPTIONS UNDERLYING THE CONCEPT FOR OVERVIEW SPACE GOVERNANCE[6]

Here are some of the assumptions that we must consider when we plan a governance system for space:

1) There is no existing governance system on Earth, nor has there ever been, that is adequate to the needs that we anticipate in space;

2) Many pathologies of human nature on Earth constitute significant obstacles to be overcome for living in space;

---

[5] W. R. Ashby (1957) *An Introduction to Cybernetics*, New York, Wiley.

[6] The Policy Sciences life works of Professor Yehezkel Dror, the Hebrew University of Jerusalem, provide a major intellectual resource for this chapter's normative space governance model, especially his Chapter 4 in *Beyond Earth: The Future of Humans in Space* (2006) and his *The Capacity to Govern* (2001), Frank Cass Publishers, London.

3) Earth's past governance failures provide lessons that are essential to learn for the future of space habitation;

4) Leadership for human civilizations in space will need concentrated education to build the capacity for Overview Space Governance;

5) Science, including the Physical, Social and Policy Sciences, Technology, Quantum Mathematics, Management, Economics, History and Human Factors should be included in that education;

6) The basic motivations for the exploration and settlement of space should be to perpetuate and improve humanity's spiritual, ethical, and religious values, and quality of life;

7) The governance of space societies should prevent the evils of crimes against society, mass killing, genocide, slavery, warfare, and ethnic cleansing;

8) Success will depend on leadership that places top priority on achieving good for humanity, while overcoming the complexities and inherent barriers;

9) Earth-based international political, scientific, and scholarly leadership must collaborate to initiate the design of government systems for human settlements in Earth orbit, on the Moon, on Mars, and throughout the Solar System.

OVERVIEW OF SPACE GOVERNANCE CHARACTERISTICS

Since space is completely foreign to human settlement, a normative first model cannot possibly predict every eventuality that will be found there. But as with all political-social systems, the issues of political philosophy and decisions about "Who Gets What, When and How"[7] will be fundamental. Here is a basic set of fundamental issues.

1) A system of leadership selection based on quality, experience and wisdom will be needed to assure moral, capable, and intelligent decisions in the framework of a firm commitment to the pursuit of agreed-upon goals. Motives of greed, corruption, or special

---

[7] Harold Dwight Lasswell, (1950) *Who Gets What, When How?* P. Smith.

interest bias cannot be accepted in any form. Moral leadership will be a critical characteristic.[8]

2) Global participation in the design of the system is mandatory, as no single nation has the resources for this massive endeavor, nor the right to represent all humanity in space. The assumption that the overall goal is for the benefit of humanity precludes control by any one nation, or by any elite group of peoples, nations, or even corporations. Global collaboration will therefore be essential, with the best current models the European Union and the International Space Station.

3) Brainpower aggregation through international Think Tanks and Study Groups will also be essential. Many of these already exist for space endeavors, but there is as yet no global study enterprise focusing on human settlement in space. This will require innovative governance and research structures to be created on Earth for planning of the 'Space Migration Epoch.' The United Nations Economic, Social and Cultural Organization (UNESCO) is the closest entity in existence today with a Mission Statement related to this task, but it will nevertheless need to be adapted for participation in the Space Settlement planning.

4) Positive human relations will be essential. On Earth, those relations have serious flaws. Inevitably there will be many kinds of families, clans and groups. A basic ethics and a do-no-harm political philosophy will be necessary.

5) The goal to occupy space and live together in a spirit of cooperation, respect and peace needs to be prescribed by a Code of Ethics.[9]

6) Citizens who have an enlightened view of Public Affairs will be necessary to achieve a functional democracy, which will require both a better understanding and a higher participation in Overview Governance than has existed on Earth.

---

[8] See Ben A. Maguad, and Robert M. Krone, "Ethics and Moral Leadership: Quality Linkages" in *Total Quality Management*, Vol 20, No. 2 (February 2009, 209-111).

[9] See K.T. Connor, Lawrence Downing and Bob Krone, "A Code of Ethics for Humans in Space." Chapter 17 in Bob Krone, Ph.D., Editor, (2006) *Beyond Earth: The Future of Humans in Space*, Apogee Space Press.

7) History shows that dictatorial and authoritative governments which do not progress according to the will and consensus of those governed eventually fail. So can democracy succeed in space? It certainly has weaknesses, as when paralysis results from conflicting desires, or when catastrophic events require dramatic actions that it cannot respond to quickly enough. Hence, an innovative political philosophy, structure, decision-system, legal system and enforcement tools are needed, a system which might be labeled "Overview Democracy."

8) Space governance presents an unprecedented challenge for global leadership. Our thinking needs to advance from national, to global, to Solar System in scope.

CONCLUSION

The simplistic concept of a freer, safer, and more prosperous community in space is too abstract, and perhaps too naïve to be credible. Dr. Jonas Salk, America's famous microbiologist, gave us a framework to think about this in his 1973 book *The Survival of the Wisest*.[10] Jonas suggested that society is in transition from Epoch "A," characterized by Darwinism's survival of the fittest, to Epoch "B," characterized by survival of the wisest. He portrayed these two epochs as Curve "A" and Curve "B" in a larger sigmoid curve, and suggests that society is now at the point of inflection. He knew Epoch "B" would be radically different from Earth's history, and he stated:

> *"The choices which Man makes from the alternatives available to him will profoundly influence his own evolutionary destiny."*

Society has indeed made progress toward the Survival of the Wisest Epoch. Democracy's Social Contract emerged in the 18th Century; slavery was rejected in the 19th Century (although remnants still exist as we begin the 21st Century); science and technology have made much of Earth's population better off; and the "Learning Organization" concept has spread through private and public entities with the Quality Sciences playing a huge role following World War II.[11] The Information Age has led to revolutionary changes in communications and the speed of business and

---

[10] Jonas Salk. *The Survival of the Wisest,* New York. Harper and Row, 1973.

[11] For a collection of *"Quality Classic Essays"* that summarize the fundamental themes, ideas and tools of the Quality Sciences, see Bob Krone, ASQ Fellow Member, www.asq711.org.

human transactions. But at the same science and technology progress has raised the probability of human extermination if wrong or evil choices are made, showing us that the transition to Epoch "B" is by no means assured as of 2009.

In aggregate, while the Benefit-to-Cost ratio for human settlements in space will be positive and enormous, to an extent that is probably beyond today's imagination, we must nevertheless do our best to understand what these benefits might be. Hence, The Kepler Space University, of which I am provost, is imagining and sponsoring a "Peace on Earth from Space" global research effort, and indeed the urgency of the human predicament and hope for space/Earth peace is captured in the University's new curriculum, concentrating on innovative solutions for humanity's needs.

Overall it is apparent that the right choices must be made, and the right management methods employed, to achieve benefits that space could offer. Overview Governance is proposed as a required 'right choice' along our journey to wisdom.

•••

BOB KRONE, PH.D.

Bob Krone is a global educator, author, and consultant in Advanced Management theory and practice. He is the Provost of Kepler Space University (2009), an Emeritus Professor of the University of Southern California (USC) in Los Angeles, U.S.A. (1975-1993); a Distinguished Visiting Professor in the School of Business at La Sierra University in Riverside, California, U.S.A. (1992-present); and an Adjunct Professor for Doctoral Programs in the International Graduate School of Business at the University of South Australia (1995-present). He has authored or co-authored six books and 70 professional journal articles.

Bob graduated from the Naval War College Command and Staff College in 1962. He is a Fellow Member of the American Society for Quality (www.ASQ711.org) with an academic specialty in the Quality Sciences; a Leadership Council Member of the Aerospace Technology Working Group (www.ATWG.org); Secretary of the National Board of the Distinguished Flying Cross Society (www.dfcsociety.org); Member Board of Directors of the Veterans Museum and Memorial Center, San Diego, California (www.veteranmuseum.org); Board Member of the Fallbrook, California VFW Post 1924; and Emeritus Member Board of Directors of Idyllwild Arts (www.idyllwildarts.org).

CHAPTER 4

SPACE CITIES, ECOMINUTIAE, & SERENITIES
MEDITATIONS ON CIVILIZATION BEYOND EARTH

PAOLO SOLERI, PH.D.

Such remarkable phenomena as life and consciousness cannot be confined to a solitary bit of matter the likes even of our astonishing planet. Hence, after four thousand million years on Earth, we are on the threshold of the space age. Animation of space then becomes a paramount consideration – the radiance of Mind reaching beyond the confines of planet Earth.

The subject can be expressed in two ways:

1. It is feasible, and thus we might want to do it.
2. It is necessary, and thus we must make it feasible.

In the first way there is no imperative, there is a drift. We might want to do it because it is challenging, escapist, amusing, work making, interesting, informative, dangerous, thrilling, etc. It has little to offer for man's dignity and his pretences of love and care, compassion and grace, it fills nicely the hedonistic paradigm of life: life is cheap, so better to have a ball whenever the opportunity arises.

In the way of the second, however, there must be an eschatological motivation, that is, a motivation derived from consideration of the ultimate end of the human journey, and therefore a command: thou shalt create, that is to say thou shalt make the improbable into the real, and thou shalt do the impossible.

In other words, thou shalt transcend thyself. Well this is an old story, 4000 millions of years old, and a success story at that.

If it is necessity rather than feasibility that demands the probing of space, then it is imperative that the space probe be developed, that is, become feasible.

But creation is not a flick of a switch. It is an eons long process in a vector matter-mind. In it, the lust for life has to grow boundless, since boundaries will keep inequity rampant.

•••

In their first generation, cosmonauts and astronauts are engaged in perfecting the technology of getting out there and, once there, survival and return.

The second generation will see the development of micro-environments such as the International Space Station, and the beginning of production cycles justifying in part the enormous cost of space ventures. It will be the age of techno-stations and avant-posts in space, and on planets and moons. It will be the foothold age.

The third generation in the space venture will be the establishment of modest ecosystems in points of specific interest in space, space cities. Because outer space is inimical to carbon based life, if and when we move there we will do so as interior designers. As any 'godlike' creatures, we would have to consider the production of modest but self-reliant, micro biospheres, i.e. ecological minutiae, hence the name ecominutiae.

But if instead we were to send not ourselves, but silicon agents to represent us, then no ecological minutiae would be necessary, just boxes, containers of minimal size, stuffed with microchips, conduits, nanoengines, etc.

Indeed, the silicon revolution has opened three possible paths. It will support the presence of humans in space, and it will also support the silicon presence, as representatives of humans in space. Chances are that both will be pursued together with the third option, cyber-hybrids, siliconized carbon man, or carbonized silicon 'man.'

Astronauts and cosmonauts speak often, and in awe, of our beautiful and serene planet. We know better. And yet this nostalgia for serenity can be transformed into intent, yielding a not-so-serene planet sending off serene ecominutiae to test and tease the mistero tremendo.

ECOMINUTIAE

The possibility of ecominutiae is pure guesswork, because no one knows what will really happen. They are thus more of a moral play than prediction, anticipation or simulation. And they join, for what they are worth, the grand dramaturgical act spellbinding man in search of himself, a creational act among so many others.

Small ecological pods, each Ecominutiae envelops a human (carbon-silicon) settlement, a city in space. Each equipped for indefinite journeys into space. For the sake of physical self-reliance and mental-biological well-being, Ecominutiae cluster together, three, four or more traveling in tandem, trading with each other, developing new technologies, new cultures, new experiences. The Serenity label is to point to a human condition that seems to escape us on this planet engrossed in its possible descent into dull materialism.

Image courtesy of the Cosanti Foundation.
Photo by Ivan Pintar. Reprinted by permission.

Given the environmental void surrounding them, ecominutiae are necessarily interiorized environs not unlike placental bags and eggs. The reverse is true of planet Earth, a sort of inside-out egg, not forgetting that this is made possible by the atmospheric blanket enveloping it.

The city in a bubble is my anticipation, a bubble that emulates a bit of our Earth. A space city, an ecominutiae, because it is a stripped down and segregated life, moving in the infinite blackness of the cosmos, desperately needs the urban effect.   That is, a city must express the interlinked paradigms of miniaturization, complexification, and duration which embodies the cell, the life form, and the urban experience as well, through which mind has emerged in evolution. Absent the construction of cities in which the urban effect is present and resplendent, life 'out there' would turn into penitentiary experiences, grim and technocratic.

Some characteristics might be observed about the design of ecominutiae:

An atmosphere will be needed, and therefore it will be created.

Gravity will vary depending on the situation.

Muscular strength will be less necessary, as will Earthlike, gravity-resistant structural systems. Further, one aspect of machismo comes from gravity. Pumping iron, the present-day muscular icon, is the triumph of muscle above weight (gravity).   But space will make machismo and pumping iron slightly ludicrous. The incongruence of bulging flesh and nothing to crush, carry, or crash with it is a fact of space. It will also be so for architectural machismo: the cantilever syndrome is in effect, the possibility of graceful structures not bound by the limits of terrestrial engineering.

But membranes will be absolutely essential. Humans will become membrane experts, or perish in the attempt.

And magnets and various forms of anchorages will be greatly useful for moving around and for various operations. In the spatial ecominutiae would be present a fractional gravity (via spinning?), and therefore legs would still be useful. Biped man is advantaged vis-à-vis the legless one who is suggested by the total absence of gravity.

Needed:

Some scientific breakthroughs,
Many technological breakthroughs,
Some socio-biological breakthroughs,
Some economical and financial breakthroughs...

And a grand noble leap of self-awareness for the whole homosphere, the leap into the serene conversion of planet Earth into the esthequity (the esthetic equity) of the cosmos.

The possible nemesis of this vision, however, is liturgical banditry, i.e. animistic crowds attempting the reconversion of the godless clusters, home of the serene biped, to Earth's ancient religio-spiritual institutional mores.

The tangible functions of ecominutiae will be:

1. Mining–processing–shipping.
2. Production.
3. Research.
4. Testing.
5. Urban laboratory, including the study of environmental and socio–cultural settings.
6. Station–vacation–resorts for space travelers, for regeneration and restoration.
7. Building, maintenance, repair and cargo stations for the space machines.

SERENITIES

Venice, in its times of splendor, was known as the 'republica serenissima,' the most serene republic, its own refined Renaissance culture, a quasi-bubble of serenity poised beside the Mediterranean seas.

Modest copycats of sorts, ecominutiae would sail weightless in the darkness of space, luminescent, chromatic, and serene.

It might turn out to be highly desirable that the ecominutiae are assembled in clusters. Then each one can 'reflect' on the others to provide reassurance, variety, safety, companionship, social and cultural exchange, technological and scientific cross-fertilization in a sort of necessary opportunism shot through and through by the imperative of the I – thou survival – interdependence.

Two main reasons for the clustering of ecominutiae are physical survivability, and psychological survivability.

The bang-bang virtuality of the media has become the dramaturgy central to the entertainment world, but it is no longer funny. Grand scale simulation of destruction, enormous scenarios of sequential physical catastrophes. A slow degrading of humaneness is going on under our very eyes and ears. And we seem to be insatiable for it, the theology of despair, monstrously violent. The Serenities' virtualities propose the opposite: the dramaturgy of optimism and of creation.

This helps the inhabitants to cope with and overcome catastrophe and the bleak darkness of space: ecominutiae congregate with others to compose Serenities, clusters of cities sailing together to 'the ends of space.'

Such clusters could 'serenely' travel for years, for generations, to other galaxies.

Serenities, a post-NASA phase in the space venture, are proposed as 'stations' aligned along the space routes that might be most convenient for space exploration.   They will be strung through space along the most convenient or the most promising routes to this or that point, this or that planet, moon, or asteroidal cluster.

To travelers deprived of sensory stimulation (sight, sound, smell, taste, touch), ecominutiae will reveal themselves as the architecture of interiors, the only possible architecture of space where the outer is life negation and the inner is life sanction.

To travelers constrained for months or years in extremely contrived spaces, such eco-stations would offer a necessary break in an environment not too foreign to their Earthly experience.   For ecominutiae residents, scientists, scholars, etc., the arrival and departure of travelers would be the most appreciated break 'of rituals,' and a connection with the far-away home planet.

Politically engaged together, the many Serenities become themselves the polis, developing knowledge, science, technologies, their own cyber-biology as space societies, each with its own culture.   They begin their evolution in the immense bubble of the cosmos, the seat of new civilizations.

THE ESCHATOLOGICAL IMPLICATIONS

The eschatological implications of 'space colonization' are fundamental and critical. Eschatology, the study of the final or ultimate end, in this case refers to the endpoint of humanity, the ultimate expression/creation of humanity in the universe in all of its spiritual, material, civilizational, and psychological dimensions.   It could be considered under the three following, but not necessarily equally important, titles:

1. The eschatological concern. The question itself of ultimate aim, the purposefulness of life, that is to say, the eschatological paradigm as such.
2. The genetic concern. The splitting of the human specie into 'sub' species as a direct consequence of a space 'invaded by humanity.'
3. The urban concern. The space probe is the urban probe on 'new grounds,' the venturing ecominutiae, over which the urban question looms ever larger in the Earth-bound destiny of the species.

With the space venture, the bulk of matter necessary to support each individual's living requirements would dramatically shrink. Instead of an individual's proportional allocation of Earth's mass, ecological surface veneer, and solar energy, approximately one-seven-billionth per person, it would be as if the Earth or any other celestial body were peeled off into successive skins (the Earth, as the last to be so treated, since the biosphere is the 'original,' unique, precious, and beautiful), each one containing an 'interiorized landscape/cityscape' of minimal physical bulk.

This is a characterization of the human environment as exponentially frugal, thousands upon millions of hollow worlds, inner-oriented worlds because of locally produced gravity, which would invade the universe from as many points of it as there are consciousness centers similar to Earth. Eventually, each galaxy would have the carrying capacity of thousands of millions of consciences (as the Earth today) billions of times over, a true explosion of consciousness throughout the physical universe.

Complexity and miniaturization will be pervasive and imperative for the mechanical, the physiological and the organizational. In other words, frugality will be the mandatory icon among all the perhaps lesser ones.

The Evolution of the Human Species

One of the most momentous outcomes of space colonization successfully carried out will be the appearance of human mutants who will more fittingly function and perform in not just one new environment, but in a variety of new environments, for we are, every one of us, children of our environment.

A most critical area of decision (or non-decision) is, therefore, the genetic pool, and with it new conceptions of human values, justice, equity, hierarchy, fairness, etc. Are we able to withstand the thought that quite possibly, I would say inevitably, the human kind might become fractionalized (fractured?) into, for instance, living fossils (the Earthlings?), plus psycho-techno man (cyber), plus superman and superwoman (an intellect-relayed multitude constituting a single creature?) similar to but on a different platform of evolution from insect colonies?

Since it will eventually not make sense to try to copy the Earth's environment in space, the congruous environment made by the space dweller will eventually define new morphological parameters for the dweller to conform to. Eventually the flesh as we know it, and as we are it now, might have to go. But well before such definitive dismissal of the organic we will probably mutate ourselves into cyborgs of one kind or another.

What, besides the ability to think, will make space humans like Earthlings? And what if the power of the thinking apparatuses becomes extravagantly different?

Since the future is not a process of deployment like the unrolling of a (sacred) scroll, but instead a process of creation, nothing that we might conjecture or plan for will ever turn out to be the future. Therefore, while it is quite possible that we will never leave the Earth, it's also quite possible that a polarization on this planet might eventually force upon us some 'strange' variances from the human kind that we are now accustomed (or resigned) to. That is to say, a genetic schism might be not only a consequence of, or peculiar to, a space colonization, but could be in fact the cause for a space colonization.

As living and thinking organisms, cosmic explorers will be subject to stresses far more critical than the technological ones. Those are assumed to be successfully cared for. In other words, the limits will be defined by the bio—technological, and by the logos–technological resilience of the explorers, especially if they are in the process of becoming dwellers of Space Cities.

As 'machines for living,' Space Cities have to sustain the onslaught of non-environment, the endless night of space, at all levels of performance.

There are three possible situations:

1. Space habitats free floating in space.
2. Space habitats grounded on planets or moons.
3. Space habitats partially or totally burrowed into asteroidal chunks of mineral matter.

The third may turn out to be the most feasible and the least expensive, besides being very promising for mining, research, learning, flexibility, mobility, etc. In any case, those habitats will also be stations, caravansaries of the space age, where the space traveler will find a less contrived environ–staging, reminiscent of the Earthly landscapes, and the beginning of autonomous local ecologies.

SPACE FOR PEACE

A. The space venture, as an urbanization of the cosmos, is tantamount to seeding it, via and with, Mind. How much of this is already in process throughout the cosmos is one of the compelling unknowns.

B. The "seeds" will be placenta/egg-like envelopes containing and nurturing the embryos of new societies. The placenta/egg reference is to be taken at the Apollonian post-bio–organic level, that is, 'designed according to Mind.' The pseudo-organic model, as proposed in some Star Wars movies, is to be carefully avoided, a life-depleting fraud.

C. For a long time, the new urban systems will be less complex, simple-minded sisters of Earth-bound towns. They will be tentative openings toward new dimensions of space–time.

D. In them, because of their very limited dimensions and complexity, human consciousness will have to cope with the most segregated and potentially deadening situations.

E. In them, given the technologically overpowering presence, the pursuit of knowledge might have as bedfellows rampant technology and exotic hedonism.

F. Since the space environs will be radically different, genetic alteration (both random and purposeful) will be only a question of time, with Silicon Mind at the outer edge of such metamorphosis.

G. The presence of flag-waving and chauvinism will be pernicious and in need of abatement. For socio-cultural obscurantism would be the result of chauvinism, Star Wars then the game of choice.

H. Technocratic space cities would be as deadly and demeaning as would be technocratic cities on this planet.

I. The effect of living in the "midst of infinity" will be an enormously confined life: the placenta/egg/embryo syndrome.

J. The eventual journey back to Earth will be cause for celebration and merriment. Upon return, the Earth will be finally perceived as the prodigious reality that it is, in its grandeur of land, sky, water, climate, flora, fauna and mind: something to behold in awe, gratitude and reverence.

Summary: So Why Space?

One could suggest two composite reasons:

1. To find out how prodigious is this planet one must try to survive and to grow into better selves outside of it. Then true reverence for Earth might set in.

2. The prodigies of planetary reality are a stepping-stones for things of an even higher degree of 'prodigiousness,' such things imply the inclusion of 'other places' in the exalting journey. Sooner or later parochialism becomes narcissism; narcissistic prodigy are not at the top on the desirability scale.

But all told, life out there in the black endlessness of the cosmos is going to be taxing for Earthlings accustomed to the endless variations, the never-ending transformations, the violences and the tranquilities of this planet.

To be drowned into a never-changing, black, inimical emptiness of the interstellar space will be a violence upon the senses and the psyche that has no parallel. Gazing into the stars as the only relief.

Therefore, the ecominutiae imperative of form and color, of light and translucency, depth and prospective. Design of the aesthetics no less indispensable than engineering of the envelopment.

Looking backwards, we see gravity has been perhaps the indispensable discipline for the development of life. Now, or soon, it might turn out to be no longer necessary or beneficial. The mega-system of the cosmos, planets and all, no longer indispensable nor necessary for life, life itself having emerged. Hence, convergence-prone gravitational fields dislodged by the supremacy of the radiant mental field.

We might then want to remember that gravity is the most unforgiving ancestor. It yet resides in and commands each cell, each molecule, each organism, each collectivity, each object. Under its weight, life moans and groans and pays a fantastically heavy tax. This is why space will turn out to be the frugal venture par excellence: it proposes the liberation of the mind.

So why go out there in the first place? If it is possible for life to free itself from the Earth, isn't it then a 'mortal sin' not to do so? And doesn't the fact that life could not even conceive of leaving the Earth before the appearance of consciousness, and that thousands of years ago the first step toward this leaving expressed itself in metaphorical, that is religious form, constitute a proof of sorts that it is the task of life, and specifically of consciousness, to do just so?

And as an 'ultimate' (eschatological) answer, because we will eventually have to nudge reality toward meaning. Along that route, we could eventually hope for a universal, that is, cosmic intellection. It is a condition promising the advent of Meaning and the resolution of inequity, in fact an esthetogenesis of reality, a transformation into grace, the singular point of grace.

PAOLO SOLERI, PH.D.

Born in Turin, Italy on June 21, 1919, Paolo Soleri was awarded his Ph.D. with highest honors in architecture from the Torino Polytechnico in 1946. He came to the United States in 1947 and spent a year-and-a-half in fellowship with Frank Lloyd Wright at Taliesin West in Arizona, and at Taliesin East in Wisconsin.

In 1956 he settled in Scottsdale, Arizona, with his late wife, Colly, and their two daughters. Dr. and Mrs. Soleri made a life-long commitment to research and experimentation in urban planning, establishing the Cosanti Foundation, a not-for-profit educational foundation.

The Foundation's major project is Arcosanti, a prototype town for 5,000 people designed by Soleri, under construction since 1970Located at Cordes Junction, in central Arizona, the project is based on Soleri's concept of "Arcology," architecture coherent with ecology. Arcology advocates cities designed to maximize the interaction and accessibility associated with an urban environment; minimize the use of energy, raw materials and land, reducing waste and environmental pollution; and allow interaction with the surrounding natural environment.

A landmark exhibition, "The Architectural Visions of Paolo Soleri," organized in 1970 by the Corcoran Gallery of Art in Washington, DC, traveled extensively in the U.S. and Canada, breaking records for attendance. "Two Suns Arcology, A Concept for Future Cities" opened at the Xerox Square Center in Rochester, New York, in 1976. In 1989, "Paolo Soleri Habitats: Ecologic Minutiae," and exhibition of arcologies, space habitats and bridges, was presented at the New York Academy of Sciences. Most recently, "Soleri's Cities, Architecture for the Planet Earth and Beyond" was featured at the Scottsdale Center for the Arts in Scottsdale, Arizona. His work has been exhibited worldwide.

Soleri has received one fellowship from the Graham Foundation and two from the Guggenheim Foundation. He has been awarded three honorary doctorates, the American Institute of Architects Gold Medal for Craftmanship in 1963, the Gold Medal from the World Biennieal of Architecture in Sofia, Bulgaria, in 1981, and the Silver Medal of the Academied' Architecture in Paris, 1984. Soleri is a distinguished lecturer in the College of Architecture at Arizona State University. He has written six books and numerous essays and monographs.

CHAPTER 5

ON THE DOMINANCE OF THE METASPHERE
THE VISIBILITIES AND INVISIBILITIES OF THE ECOLOGY/EXOLOGY OF SKY AND SPACE

LOWRY BURGESS, PH.D.
DEAN, PROFESSOR, CARNEGIE MELLON

I. INTRODUCTION

Each generation's most imaginative intuitions concerning the meaning of human BEING are formed and dominated by its concept of the sky, and the cosmos beyond. It is there in the overarching vastness that mythic logic – the logic that seeks to grasp and frame the unknown – plays out, and finds initial expression, and then gives birth to the cultural frameworks of society. Each age has its own inward hopes and its own utopian aspirations that are often envisioned in outer space, the place beyond. It may be a transcendent place, heaven, or a place that illuminates cultural belief frameworks, mythology. These frameworks then take the form of art and language, morals and ethics, science and technology, economics and politics. But they originate there, as we look outward, where we see our inner selves reflected back to us.

What do we see as the ultimate goal for human consciousness in the cosmos beyond universe? And how will our cosmic adventure relate to humanity's destiny here on Earth? To undertake such an exploration, ontological and teleological, raises all the questions most essentially human. As we reach outward, an inward space equally yawns open, disclosing inner frameworks of mind/body, culture/society, ecology and history.

II. THE METASPHERE

I call the contemporary manifestation of our outward quest the Metasphere: a world-binding shell, a layer of humanly-created meaning comprised of policies, technologies, and energies that surrounds the entire Earth and reaches into outer space. It is a highly complex, transparent fabric and architecture that has spanned the Earth. And most significantly, it is forcefully reshaping life on Earth. It is massively shifting our view of the planet from weather and climate to new forms of agriculture, from population depredation to global health, from instantaneous global economics to cross-cultural communications, from the depth and currents of the oceans to the intimate presence of other planets and the deep heavens beyond. In other words, practically every action and interaction is redefined by the presence of the Metasphere from civilization to the wild, from the self to the other.

It is an invisible, dense 'plasma' that includes everyone in its powerfully influencing/controlling grasp, as though all the human projections into outer space have been reflected back into the atmosphere, and outer space is now at the tip of our noses: outer space is right here! The Metasphere is intimate, immediate, and yet totally indifferent.

This indifference is driven by the vast set of systems: technological, economic and political, including functioning theories and political process that are formalistic and in-grouping, even incestuous; that few if any have control or command over, let alone have knowledge of it. It is a global simultaneous Brownian field of interactions of enormous power.

This dominating indifference is forcefully inverting all human relationships, as if we are walking upside down with our bodies turned inside-out. Previously, all relationships to value were Earth bound, territorial, horizontal, gravitational and basically substantial, locatable in time and space. Now all value and power is invisible in the sky/Metasphere, everywhere and nowhere simultaneously. The Metasphere's 'gravitational' power is so great that our feet are metaphorically in the sky!

At the same time the 'self' is so globally distributed that the far is near and near far – the wilderness and the unknown are within – the self afar. These fundamental and forceful displacements can be seen in contemporary economic/political/cultural events.

As in ancient times, when the stars and sky determined human destiny, so too now the rapidly evolving Metasphere constitutes a mostly invisible and highly intangible outer ecology (ergo, an exology) of Sky and Space. It wraps the Earth in an invisible binding, driving unstated destinies; ruthlessly pursued amid blatant contradictions and hostilities. The Metasphere's ruthlessness is driven by its own formalistic self-containment and the inability to necessarily control it or place it in reference to anything else other than its own evolutionary energy. The unknowability of its teleology leads us to the question of the destiny -- where it will drive or pull us? Into this transparent Metasphere the human mind looks to understand its deep past and its deep future.

All of what is stated above has been influenced over the past 50 years by my readings and sometimes contact with an amazing group of contemporary intellects. Most directly notable are theologian Teilhard de Chardin's concept of the noosphere, biologist Rupert Sheldrake's concept of morphogenic fields, McLuhan's sense of media culture, Soleri's model of eschatology, and Martin Heidigger's concept of gelassenheit or 'releasement.' Looking back in time I draw from Plato's Timeaus and behind Plato's Socrates to the frameworks and topologies of Egyptian thought and religion.

Parallel to all of these are Santillana and Von Dechend's encyclopedic "Hamlets Mill," and the vastness of human sky mythologies as conveyed by Joseph Campbell, as well as studies in Buddhist and Hindu cultures and other anthropologies. More recently, I have been directly involved with the inherent ideas that are forming in the internet, the free-ware and shareware phenomena, and their larger meanings in relation to shared global intellectual property that is of enormous economic value. All of this is paralleled with an obsession and visits to archaeo-astronomical sites around the world.

III. THE METASPHERE'S DRIVING FORCES

Today, there is no conflict anywhere in the world that is not aggravated by the domineering presence of the Metasphere, and therefore it is as essential to understand the Metasphere's inherent topologies and its interactive 'exology' as it is to study Earth's climate and ecology.

A few of the major forces driving this Metaspheric concatenation are:

1. The emergence of the computer, combined with global communications systems.

2. The emergence of new global bodies of laws, accords, and policies.

3. DNA and genetic engineering, yielding human control of life forms.

4. Brain re-forming chemistry through psychotropic drugs that can change the aging of the brain, altering mind/body relationships.

5. Internal and external robotics, the micro/macro machine.

6. Sustained Earthly life beyond Earth.

7. Life as a universal presence.

8. The new infinite cosmology, the official scientific shift to an infinitely ever-expanding model of the universe.

The highly intangible ecology of Sky and Space of the Metasphere is a dynamically interacting holism that is barely visible, rarely studied, and hardly discussed, other than within the fragmented domains which constitute its roots.

The Metasphere is inhabited by an invisible and secretive topology of the amalgam of the electronic-digital-pixel-fiscal-security-network called the ~digipix-plasma~. This ~plasma~ is the electronically mediated atmospheric ecology of 21st Century. In this futuristic ~archaeology~ the rules and dynamics of isolated and dislocated consciousness of the dystopia called the 'Where-Nothing-Happens' ~ are revealed. Yet within this inexorably grim topology are glimpses or flashes of another ~agri-culture~ that inhabits that reality.

It nearly goes without saying that electronic data is converging into a digipix-plasma (the digital-pixelated, data/icon-amalgam), available anywhere at anytime. We will need no longer to be tied to networks or lines, and data is no longer linear in form but rather a plasma-weather-system, gas-like structure, that we inhale and exhale and depend upon as the air we breathe. Just as in a snowstorm or the drift of fog, all things become blurred, just so, in the ~digipix-plasma~ everything melts into a single whiteness - a white noise. Its mistiness secrets everything; hiding even itself in an atmosphere of secrecy.

The rules that govern this topology are more akin to a mix of hydrodynamics, a really good mystery, and Earth tectonics, as well as those baffling rules of a complex and perplexing dream. With variable technologies, we surf, scan, and click across the digipix-plasma wherein the broadcast media, communications technologies, internet, global economics system, and world-wide security systems are becoming more and more convergent. We live in the nearly invisible self-secreting ~snowy~ swirl of digipix-plasma. This simultaneous plasma of global media brings confrontational cultures into violently contesting immediacy that was in the past isolated by the distances of time and space; and thereby changes most of the basis of typical socio-economic-cultural relations. Finally, this leads to the underlying frameworks of emotions that inhabit the Metasphere, and the crucial task is to reframe its frightening power into new insights and new expressions of utopic rather than dystopic projections.

Aesthetic invention in the Metasphere fashions and explores a new vastly expressive, intimate, synaesthetic multisensory consciousness. Now, in the Metasphere, there has evolved a vibrant mesh directed inward by a tremendous psychic gravitation toward the goals of intense integration of the senses in synaesthesia, combined with the coherence of 4D consciousness. This mesh is somewhat like static electricity unexpectedly jumping. It is stochastic in its behavior and nearly convulsive in its effect, often producing astral or out-of-body effects with attendant ecstasy and fear. This communicative, interactive Metaspheric 'garden' constitutes its own new structures and contents. The pre-existing image wants to be dynamic, more rapid, more explorative, and inquisitive; it wants to be more democratic, more synaesthetic, more polyvalent, more free-associative. In other words, this new framework demands a de-gestalting though fragmention, and to then be re-assembled on another plane beyond the Metasphere. It requires entirely new types of image formulations which are dynamic, hieroglyphic and synaesthetic in nature, and iconic in meaning – a new utopic projection.

To allow these frameworks of the Metasphere to continue unabated and not understood is to strangle the life-blood of the imagination and of our future reaching out to the cosmos, because inherent in the Metasphere are control and power archetypes wrapped with the various illusions of participation and access.

Having said all this, it obviously becomes imperative that we gain mastery within the Metasphere to search out its positive potentials that can then loop out in a further expressive expansion outward again in the coming generation. This new understanding will require that networks of university, institutional, and space agencies cooperate to create an integrated view of this newly domineering by-product of humanity.

New policies and actions can lead to better understanding and to positive future developments in an emerging framework that links, interconnects, and unifies an extraordinary potential for creativity within networked groups, institutions, and facilities around the world.

So let us call upon the world's space communities, its space agencies, researchers, academics, lawyers, economists, artists, and institutions to extend cultural and interdisciplinary hospitality toward each other, to engage in understanding the emergent Metasphere with and for the whole of humankind.

IV. HUMAN CULTURE AND THE METASPHERE

In the last century, the large elements of human Culture, including values, language, art, science, and technology have tended toward self-isolation and solipsism. Consequently, each moved in its own gravitational field, with enormous internalized momentum. Today, the struggle between the contradictory desires reflected in diverging expectations has resulted in confusion and traditional cultural stagnation.

Viewed traditionally, the construct of Culture reveals difficult interactions between the massive and powerful domains of Art, Science, Technology, Language, Values/Beliefs, and Religion. Each of these domains had created a world unto itself, responding to its own inward needs and developments, interacting only awkwardly with the others, and, in many cases manifesting overtly destructive or repressive tendencies. Each had its own set of sub-categories and relationships that made each a complete, self-defined culture within itself. For example, science had its own art, its own terminology, its own technology and techniques, its own science, and even its own 'religion,' as did art, technology, religion, etc.

Interlinking these distinctive cultures are three fundamental vectors, vast assemblies of inter-related ideas, technologies and techniques that transform 'modern' and 'post-modern' relationships by inverting and reversing them, and thereby create new relationships. They are:

Techne, the combination of science, technology, and technique, as manifested most recently in the digital revolution.

Ecos, the integration of planetary systems knowledge, which views the planet and the cosmos as one living, intensely-interrelated whole system, reaching from DNA to the life-strewn cosmos.

Mythos, the quest for a structure of knowing, of meaningful existence, of belonging.

Techne, Ecos, and Mythos are all animated and furiously driven by an overpowering information and communication media burst within the Metasphere. All are fractiously interacting with each other, and the resulting enormous friction has immediate social and environmental consequences, some of which are explored below.

V. MANIFESTATIONS OF THE METASPHERE

A. LAW, POLICY, AND THE METASPHERE

From the broad idea of human rights, stemming from ancient customary law and articulated through the Roman concept of 'ius gentium,' has grown a vast body of commonly agreed upon laws, accords, and declarations that have become an interlinked canopy of agreements that have been endorsed by most nations of the world. These constitute the legal and policy framework for the survival of humankind and our environment, and our emergence into a more secure and sustainable future.

The convergent moment at the end of World War II was the United Nations Declaration of Human Rights, further articulated in the Geneva and Hague Accords, and augmented by Antarctic Law and environmental policies, Sea Law, and now the outward reach of Space Law. This dense set of shared relationships, precedents and mutual protections for peoples, cultures and environments, including the protections for cultural history, has emerged as framework I call the 'Common Code,' an implicit legal framework that unifies the world in an evolving effort toward expanding security and justice.

Although the Common Code is not widely recognized, it constitutes a framework of globally impacting precedents that can become the stable foundation for growth of new global relationships, as well as nation-building procedures.

Still, grave tensions beset this framework. For example, the expansion of NATO to include Scandinavia as well as nations in the Southern Pacific Rim has recently been discussed; those geographies are critical to the internet and geo-satellite networks. Is NATO defense policy expanding to include the communications frameworks of the internet? And does this explain the intense reaction to the recent Chinese missile destruction of one of their own satellites in outer space?

B. GLOBAL FINANCE AND THE METASPHERE

At any moment, around $400 trillion in various forms of value is being exchanged through globally interconnected financial systems. At least 30 separate major forms of value and types of commodities interact in this huge dynamic flow between the Federal Reserves, the World Bank, and the International Monetary Fund, with and among other connected institutions and governments worldwide. Many forms of normally negatively conceived economic value, including credits, debts, swaps, loans, and deficits are used in this system to maintain balance and mobility, and to provide further financial opportunities. In their pursuit of economic advantage, peoples, countries, and regions thereby develop different forms of value to participate within the global finance system.

The global finance system requires dynamic flow. Value must move and transform, which requires instruments such as bonds, loans, credits, deficits and debts, as much as it requires commodities in different forms, such as gold, cash, stocks, securities, futures, etc. The circulation of electro-value is so vast and so instantaneous that it inverts all normal and traditional meaning referents.

This aspect of the Metasphere, the econosphere, has more power, force and immediacy than any other humanly created structure. This exchange, circulation and interaction in simultaneous time is controlled and monitored by the huge global Federal Reserve System, with its distributed centers in major nations. All interactions must be maintained in energetic balance for it to remain invisible to the inhabitants of the world, but when cracks appear, governments fall and whole peoples are enslaved, wars break out, and fortunes vanish.

When the electro-value swirl is combined with the broadcast sphere and added to the sum of all video and computer broadcasts and integrated with global communications systems, as well as defense and security systems; all are made transparent and are interactive to and with each-other -- then we can intuit the Metasphere and we can see an entirely new volatile topological landscape of domineering power.

We can clearly perceive that this Metaspheric framework commands a new politics and economics. It will mean a still further disengagement from typical and traditional political and economic behaviors. Unique forms of power will be created and used while rhetorically camouflaged and carefully guised, as seen in the many recent realignments among the National economies and their relationship to the global finance system, or in the 50 year positive gains in global health as seen from statistics from the World Health Organization that are generally unknown and unseen, or in the emerging agro-revolution that is silently forming.

C. PROPERTY AND THE METASPHERE

Within this most disengaged complex of technologies, a framework of new types of shared 'intellectual-properties' is being formed to rage against established forms of exclusive property ownership. On the other hand, free, growing and globally shared 'IP' is expanding into worldwide communications and finance systems, creating an economic 'free zone' and literal 'common-wealth.' By so doing, a new reserve of future value is established. The many examples of 'share-ware,' and mutually developed internet value are examples of new forms of globally shared properties. The current dilemmas about copyright, plagiarism, and ownership are driven by the dynamic shredding of traditional property rights, and by the anonymous instantaneous presence of the domineering Metasphere. This vast complex of forces comprising the Metasphere is transforming much of the traditional sense of ownership and property. There will be many new forms of emergent 'property' that will have global impact. One such 'property' that is of enormous value is abandoned or endangered historic/cultural sites, which are becoming protected within an emerging sense of their belonging to commonly and globally shared rights to historical memory. This idea of the emerging global Common-wealth is slowly developing a new sense of property within the Metasphere.

D. THE HUMAN BODY IN THE METASPHERE

We do not know what the human genome wishes to become. Certainly, most of the physical and mental structures that have evolved to enable the genome to function in gravity will become superfluous in the zero gravity and multiple gravity environments beyond Earth. What is it that life, and human life in particular, will become beyond Earth? Recent science and technology (with technique) have turned all bodily topological relationships inside out and upside-down. Technology has penetrated into all bodily systems, the body is penetrated and permeated with technology. The relation between the robot and the human will become increasingly transparent.

Technologically, everyone's 'body' will be disengaged and globally redistributed. Gravity-locked time becomes unfettered in simultaneous time, while at the same time the actual bodily presence will become a virtualized non-presence.

Medically, the Metasphere massively affects the health of the world through its capacity to monitor droughts, famines, plagues, migrations -- volcanoes, earthquakes, tsunamis, tornados, and hurricanes. All aspects of human health are involved in the communications/economics of the Metasphere, and its ability to make such catastrophes globally simultaneous is extraordinarily useful, from often-conflicting points of view. The knowledge of such catastrophes' and the use of the knowledge has a specific sociopolitical application, particularly with whether and how to intervene with help.

Humorously, we all have seen this new reality that embodies the disembodied and disembodies the embodied. We have seen people vanish into their 'cell' phones. Disembodied creatures with two heads, one-ear-to-cell-phone and one ear-to-i-pod, right and left arms joined, walk around in the realm of the no-private/no-public, thereby privatizing public space and publicizing private space. Everyone's DNA is in full display while chatting into the air, communing with unseen presences.

E. TECHNOLOGY/CULTURE AND ITS IMPACT ON THE METASPHERE

If we took away all the technology, networks, hardware, software, etc., what are humans doing? What global disembodiment and displacement is unfolding? We are disembodying fragments of ourselves to enter into new relationships, multi-nodal and synaesthetic intensities of simultaneous communication and communion developing an aetheric sphere -- timeless and spaceless -- an Earth disengaged from Earth.

As with any new tool-set and its emergent techniques, there is a distorted tendency to concentrate on the 'objective' impact, and to neglect its extremely complex interactions in the psyche. Every new tool-set opens positive or negative ways toward inward articulations and realizations. In the largest frame, the Metasphere, with its connected technologies, softwares, and techniques is a complex mirror that reflects and images the largest axiologies of culture. It then becomes the underlying vector driving culture forward for individual's, as well as for whole Cultures.

This combined and complex technological and technical presence, with its dazzling array of phenomena, is the latest means toward the externalization of human consciousness. This newly disengaged life inheres limits and taboos, potentials and liberties, anxieties and fears.

Most radical is the multifaceted flowing and multidirectional dynamism of the Metasphere. Its deep communicative and expressive power is being tapped virulently. We see that power in blazing flashes, both inward and outward, as if it were reflected flashes between transparent windows.

In this environment, the formal structures of the space-time experience are extremely volatile, impacting directly on the real-time presence of events, creating a condition that has never been experienced before. Event-time and network time interact: an event that is very fragmented and disconnected is smooth on the network; events which seem smooth in real-time appear disrupted and incoherent on the net. This creates a social friction that has the feeling of a single discontinuous time zone whose fractal branches fold back on themselves.

VI. CONCLUSION

With various technologies we can surf, scan, and click across the information surface, where all media are converging in an interrelated amalgam that is invisible to us. When we venture upon it we don't know what we are on or what is underneath its surface – what currents are flowing there, and at what depth the hidden monsters lie in wait! When this global electro-swirl combines with the broadcast sphere, is integrated with video and computer broadcast global communications, and is wrapped within defense and security systems, the systems become interactive to and with each other – then we can intuit the Metasphere! It is an entirely volatile topological landscape of such power that it consumes us in its data-storms and media-hurricanes. These interactive data-flows reinforce each others' objective illusions: 'if its there on TV, it must be true,' and if 'it's there on TV and here on the internet, then it must be REALLY true!'

The cultural implications of such a powerful omnipresence are fundamental. The fractious interactions between culture and economics, law and policy, environment and health, resources and capital are of immediate concern. Most acutely focused now are energy needs pitched against environmental fears, under the umbrella of 'climate change,' exacerbating socio/political tensions and large cultural conflicts. Dangerously, most of today's global conflicts are aggravated by, if not driven by, the Metasphere's domineering yet invisible presence.

So here we are, facing the critical need for holistic study and a deep understanding of the domineering presence of the Metasphere that is dynamically and sometimes violently reshaping fundamental Earthly relationships on a global scale. This situation calls for the creation of holistic interdisciplinary attention to the dynamic totality or 'exology' of the Metasphere, where all aspects of human culture interact in the sky and space surrounding the Earth.

•••

LOWRY BURGESS, PH.D.

Having been educated at the Pennsylvania Academy of the Fine Arts and the University of Pennsylvania and at the Instituto Allende in San Miguel Mexico, Lowry Burgess is an internationally renowned artist and educator who created the first official art payload taken into outer space by NASA in 1989 among his many Space Art works. He is considered one of the few pioneers of the Space Art movement that now has grown to hundreds of artists all over the world.

After the destruction of the Buddhas in Bamiyan, Afghanistan in 2001, he authored the "Toronto Manifesto, The Right to Human Memory" that received worldwide endorsement. One of the provisions of the Manifesto has led to the creation of a new global value/incentive for the protection of cultural sites throughout the world. This new value/incentive is in the process being implemented by UNESCO and the World Bank.

His artworks are in museums and archives in the US and Europe. He has exhibited widely in art and science museums in the US, Canada, throughout

Europe, as well as Japan including various internationals such as Documenta, the Vienna Biennal and his recent solo exhibition at the Carnegie Museum of Art. Art Historian Raymond Vezina, at the University of Quebec, states that "He shares this utopic, visionary tradition extending from Saint Augustine, through Dante, Thomas Moore to William Blake and the American transcendentalists of the 19th century: Henry David Thoreau, Ralph Waldo Emerson and, more recently Gyorgy Kepes."

He is Professor of Art and former Dean of the College of Fine Arts and Distinguished Fellow in the STUDIO for Creative Inquiry at Carnegie Mellon. He has founded and administrated many departments, programs and institutions during his 45 years as an educator in the arts. He has created curricula in the arts and humanities in the US and Europe while serving for twelve years on the National Humanities Faculty.

For 27 years he has been a Fellow, Senior Consultant and Advisor at the Center for Advanced Visual Studies at MIT in Cambridge, Massachusetts where he created and directed large collaborative projects and festivals in the US and Europe.

"First Night", the international New Year's arts festival, was created and founded by him. He originated the first "Arts in the Subways" program for the Department of Transportation and has developed and advised in more than a dozen major city scale projects.

He has received awards from the American Academy of Arts and Letters, the National Institute of Arts and Letters, the Guggenheim Foundation, the Rockefeller Foundation and several awards from the National Endowment for the Arts and the Massachusetts Artists Foundation, and the Kellogg Foundation and the Berkmann Fund. He received the Leonardo Da Vinci Space Art Award from the National Space Society. His book, "Burgess, the Quiet Axis" received the Imperishable Gold Award from Le Devoir in Montreal.

Among his hundreds of exhibitions and performances, most recently, his artworks have been exhibited at SETI in Mountain View, CA., the Festival of Art Outsiders, and the CNES, the French Space Agency in Paris, as well as a solo exhibition at the Carnegie Museum of Art in Pittsburgh and with his newly formed "Deep Space Signaling Group" in an artwork involving the International Space Station and NASA in April 2008. He continues work on new aspects of his lifework, the "Quiet Axis".

He has been featured in television and radio broadcasts in the US, Europe, Canada and Japan. (NOVA, "Artists in the Lab"; Smithsonian World, "Elephant on a Hill", "Artists of Earthwatch", "Arts and New Technologies" (Tokyo 12); "Artransition" (Austrian, German National Television and 24 other state television systems); "The Quiet Axis" (Hungarian State Television). He has been a guest speaker on more than two hundred national and international radio broadcasts including 3 NPR broadcasts. He has also appeared on the CBS Today Show and has made countless other appearances on television in Canada and Europe and has been widely published in numerous books, newspapers and magazines.

CHAPTER 6

SELF-HEALING SOUNDS BEYOND THE ETHERIC WEB

BARBARA HERO

We are on a space station called Earth,
one planet among the music of the spheres.

INTRODUCTION

On Earth we live in a world of duality, a world of space and time, where space and time are measured differently, one in distance, as perhaps inches or centimeters in visual art, the other in frequencies as cycles per second in audible music. In our Earth-bound setting, music allows us to live in the moment, because the duality of space and time seems to vanish as we experience music in time alone. Hence, music has been a vital aspect of human culture over the millennia, and I believe that used appropriately, it will increase the already heightened consciousness that humans experience when we live in space. I anticipate that it will help us tune into the "universal mind" that seems to be part of the space experience. Living in space would then be experienced as a condensation of the space-time duality, requiring us to relearn the laws of scientific harmony, and of audio and visual constructs.

This chapter describes my research, theories, and practices related to self-healing through music. I will describe experiments with music on Earth by telling you my story, which began when a man said to me, "That instrument that you have invented was brought to you from the future," and I will also explore what pure harmonic music might be like for those living in space.

My "instrument from the future" which I call the "Pythagorean Lambdoma Harmonic Keyboard" (PLHK) is a unique musical keyboard instrument that has an array of diamond shaped keys. It has seven colors in seven rays that radiate from the diamond-shaped matrix.

Figure 1
The Pythagorean Lambdoma Harmonic Keyboard" (PLHK)

It has mesmerized players since 1993, 30 years after I had the initial vision that inspired it. As a musician presses a key, the instrument creates stereophonic, harmonic intervals consisting of two tones, and at the same time it creates two-dimensional multi-hued Lissajous patterns on its visual display, generated by the frequencies of the intervals being played.

MY MUSIC BACKGROUND ON EARTH

As an artist, I dreamt of the paintings that I would create. I would get up in the morning, stretch a canvas, gesso it, draw the outline of the vision from my dream, and begin to paint from the center of the canvas. I created colored details of checkerboard lines representing reflections in water within circles representing bridges. In the evenings, I played Mozart sonatas on the piano.

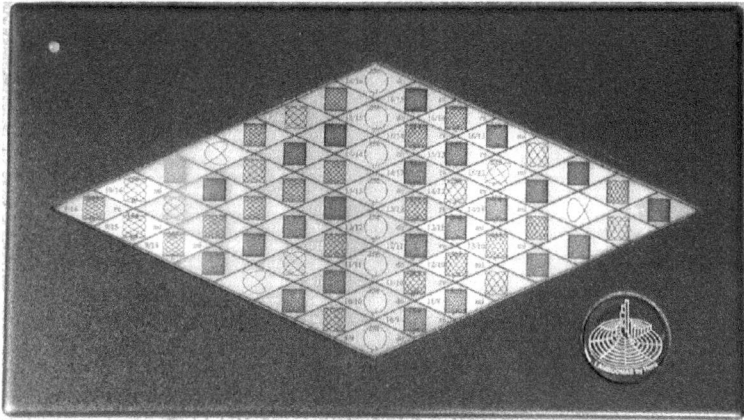

Figure 2
Lissajous Figures on Keys of 4th Quadrant (Oracle)

This daily rhythm created the urge in me to find the link between art and music, and for years I searched for it. The term "audio-visual" captured my imagination, and eventually I found the link between art and music to be the fourfold relationships of 1) ratios, 2) intervals, 3) frequencies and 4) wavelengths and colors.[1] Now I use the PLHK to simultaneously create sound and the corresponding colored geometric figures that bridge the divide.

An example of using wavelengths and colors to translate from music to art is seen in Figure 3.

---

[1] 1) Ratios are fractions 8/8, 9/8, 10/8, 11/8, 12/8, 13/8, 14/8 and 15/8.
   2) Intervals are steps of a musical scale tonic, second, third, fourth, fifth, sixth and seventh (as seen in the sequence of the ratios above.)
   3) Frequencies are measured in cycles per second indicating specific harmonic pitches in music. The numbers 1, 3, 5, 7, 9, 11, 13 and 15 are raised to an audible reference octave by doubling. The resulting frequencies in the reference octave are: 256 P(C), 288 Q(D), 320 R(E), 352 S(F), 384 T(G), 416 U(A), V448 (A+), 480 W(B) and512 P(C). (These follow the same sequence of ratios and intervals.) The colors in this sequence are red, orange, yellow, green, blue, indigo, purple and violet. The complementary colors in this same sequence are green, aqua, blue, violet, magenta, orange, yellow orange and gold.
   4) Wavelengths are measured in proportionate distances and can be applied to creating harmonic art based upon sub harmonic music, 1/1, 1/2, 1/3, 1/5, 1/7, 1/9, 1/11, 1/13 and 1/15. (This sequence represents just one row of seven rows of the sub harmonic matrix.)

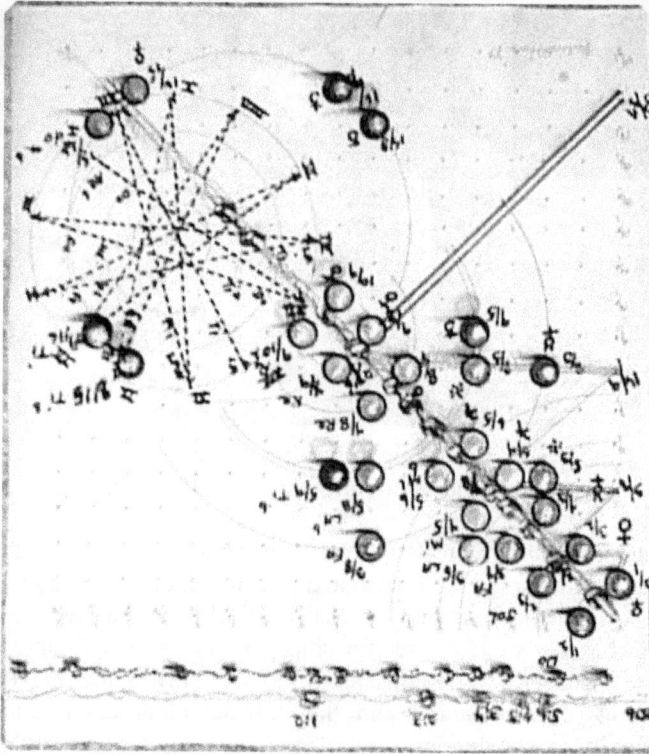

Figure 3
A Pattern of Wavelengths, Ratios, Notations, and Planetary
Symbols on the PLHK Musical Instrument

Figure 3 shows the 16 by 16 Lambdoma matrix turned 90
degrees. The string length at the right top indicates the 1/2,
2/3, 3/4 (a sub harmonic division of a string length). At the top
left of the drawing the balls representing musical notes indicate
the musical ratios going down on either side of the diagonal.
This is an example of how one might play the PLHK by pressing
each ball until coming to the circle of ratios 15/16 (sub
harmonics) and ratios16/15 (harmonics). The ratio and the
solfeggio scale, as well as symbols for the planets, are
indicated by the pattern.

"IN THE BEGINNING WAS THE WORD"

The word was sound. The word was communication. The word was song.
A grunt expresses a feeling of pain, pleasure, alarm, relief, anger, or love.
The grunt turns to vowels, ay, eee, eye, ooh, uue and finally aah. The
sound aah expresses wonderment, satisfaction and anticipation of 'what
next?' The notes do, re, mi, fa, sol la, si and do form an octave, ascending

steps of a scale, as does a, b, c, d, e, f and g or p, q, r, s, t, u, v and w. Colors indigo, violet, red, orange, yellow, green and blue are seen as the voice ascends these eight steps.

Which pitch will resonate with you as you sing up the scale? Discovering your resonant pitch is discovering your healing frequency of sound.

Beats in time ascend with the melody of the steps. A melody jumps from one note to another until a pattern forms. While playing or singing these pitches, shapes of intervals form in circles and square web-like structures. Thus, music is vibration that can be measured as ratios in time. It can also be measured in ratios of wavelengths in space using the musical laws of ratios of intervals. Lines may be drawn that are proportionate to the ratios of musical intervals, giving us a specific linkage between music and visual arts.

ART AS SELF-HEALING WAVELENGTHS THAT ARE BASED ON RATIOS IN MUSIC:
THE CONNECTION BETWEEN MUSIC AND VISUAL ARTS

The full length of a 32 inch string can be considered as its fundamental wavelength. It has a frequency of 424 cycles per second (cps), which is a musical note close to A on the traditional musical scale, and a color of indigo.

If the string is divided in half, it is still the same musical note, but it is one octave higher, at a frequency of 848 cps. It still has the same color, and a length of 16 inches.

Figure 4 illustrates a way of conceiving a musical tone system language based on an actual biological shell. The ratios numbered from 1:1 to 9:8 on the shell represent a harmonic visible world. The lower part of the shell in this case represents the invisible energy world of sub harmonics, 1:2, 1:3 ... The placement of the balls represent the rings of growth of the shell. The web to the lower right is an example of wavelengths that may be translated into musical notation based upon their different measured wavelengths.

If the string is divided into 2/3, it has a ratio of 2:3, an interval of a fourth, a frequency of 646 cps which is a musical note close to E, a color of yellow, and a length of approximately 21.33 inches.

'Keynotes' are the intervals of the key that an individual most enjoys hearing when they play the harmonic and sub harmonic frequencies on the PLHK. Musical notations that span the octave fall between the traditional keys A, B, C, D, E, F, and G, which I named P, Q, R, S, T, U, V, and W.

If the note C (P) is chosen as a keynote, it has a color of red, a ratio of 1:1, a musical frequency of 256 cycles per second, and a wavelength of

53 inches. That particular keynote of C at 256 cycles also generates its own 16 by 16 matrix field of intervals by multiplying each ratio in the field by the keynote C 256Hz.

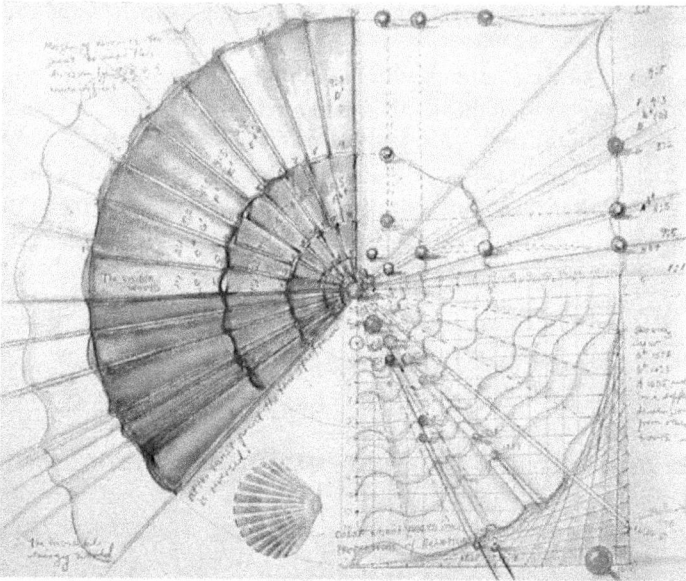

Figure 4
Art based upon the Music of a Shell: The invisible energy world.

LINKING SPACE AND TIME

Art that is based upon musical string lengths illustrates a use of sub harmonic music in art. Art (length) and music (time) seem always to exist in an inverse relationship.

A musical string can be cut into sub harmonic notes proportional to the interval of an octave in music. Sub harmonic intervals are defined as ratios in which the numerators are less than the denominators. 2/3 of that length represents an interval of a fourth, 4/5 an interval of a sixth, and 8/9 an interval of a second.

Finding a common denominator in mathematical ratios enables us to translate art into music and music into art, linking the measurements of space and time. I have found that as a form of meditation, the combination of playing the instrument and watching the vibrations enhances our audiovisual experience and inspires a physical and mental self-healing experience.

We devise a chart showing how the wavelength of sound measured in feet or inches can be translated into frequency in cycles per second.

When we determine the frequency or wavelength of sound, we can translate one into the other. For example, the velocity of sound in air is approximately 1130 feet per second at room temperature and pressure.

The formula is: V = F x W
Where V = velocity in ft/second
F = frequency in cycles/sec
W = wavelength in feet

```
M. FT. IN.    FREQ. TNE.OCT.COLOR         /M /r /r /r /   / /v,
0    1   12   1130        D  2   GR BL      25  86  1032 13.139535  A  -5  RD OR
0    2   24   565         D  1   GR BL      25  87  1044 12.965506  A--5  RED
0    3   36   376.66667   D  0   MAGEN      25  88  1056 12.840909  A--5  RED
1    4   48   282.5       D -1   GR BL      27  89  1068 12.696629  A--5  RED
1    5   60   226         B--1   ORANG      27  90  1080 12.555555  A--5  RED
1    6   72   188.33333   G -1   MAGEN      27  91  1092 12.417582  A--5  RED
2    7   84   161.42857   E -1   VIOLT      28  92  1104 12.282609  G  -5  MAGEN
2    8   96   144.25      D -1   GR BL      28  93  1116 12.150538  G  -5  MAGEN
2    9  108   125.55556   C--1   GREEN      28  94  1128 12.021277  G  -5  MAGEN
3   10  120   113         C -1   ORANG      28  95  1140 11.894737  G  -5  MAGEN
3   11  132   102.72727   A--2   RED        29  96  1152 11.770833  G  -5  MAGEN
3   12  144   94.166667   G  -2   MAGEN      29  97  1164 11.649485  G--5  INDGO
4   13  156   86.923077   F  -2   PURPL      29  98  1176 11.530612  G--5  INDGO
4   14  168   80.714286   E  -2   VIOLT      30  99  1188 11.414141  G--5  INDGO
4   15  180   75.333333   E  -2   BLUE       30 100  1200 11.3       G--5  INDGO
5   16  192   70.625      D  -2   GR BL      30 101  1212 11.188119  G--5  INDGO
5   17  204   66.470588   D--2   BL GR      31 102  1224 11.078431  G--5  INDGO
5   18  216   62.777778   C  -2   GREEN      31 103  1236 10.970874  F  -5  PURPL
6   19  228   59.473684   B  -3   YELLO      31 104  1248 10.865385  F  -5  PURPL
6   20  240   56.5        B  -3   ORANG      32 105  1260 10.761905  F  -5  PURPL
6   21  252   53.809524   A  -3   RD OR      32 106  1272 10.660377  F  -5  PURPL
6   22  264   51.363636   A  -3   RED        32 107  1284 10.560748  F  -5  PURPL
7   23  276   49.130435   G  -3   MAGEN      32 108  1296 10.462963  F  -5  PURPL
7   24  288   47.083333   G  -3   MAGEN      33 109  1308 10.366972  F  -5  VIOLT
7   25  300   45.2        G  -3   INDGO      33 110  1320 10.272727  F  -5  VIOLT
8   26  312   43.461539   F  -3   PURPL      33 111  1332 10.18018   F  -5  VIOLT
8   27  324   41.851852   F  -3   PURPL      34 112  1344 10.089286  E  -5  VIOLT
8   28  336   40.357143   E  -3   VIOLT      34 113  1356 10         E  -5  VIOLT
8   29  348   38.965517   E  -3   BLUE       34 114  1368 9.9122807  E  -5  VIOLT
9   30  360   37.666667   F  -3   BLUE       35 115  1380 9.8260907  E  -5  VIOLT
```

Chart 1
A Portion of a Chart Showing Feet, Frequency, Note, Octave, Color
(in frequencies)

The Reference Octave is between 2' @ 565Hz and 4' @263Hz. Note that 17' (the King's Chamber length) has a frequency of 66.470588Hz. and that 66' has a frequency of 17.121212Hz. They are both Db notes and aqua color and 1 octave apart.

When we take a wavelength as a dimension of a length, width or height of a room, we can find the resonant chords of the room.

This analysis was applied to the King's Chamber in the Great Pyramid of Egypt in 1984, and the resonance of the 17 ft. by 34 ft. room was found to be 66 cps for 17 ft. and 33 cps for 34 ft., a musical note of PQ (Db or C#) (10 meters or 20 meters). The musical relationship of the lengths and frequencies of the Chamber is the interval of an octave, a ratio of 2:1 or 1:2, multiplying or dividing the frequency by 2. The color of the Chamber would be an opaque orange or a transparent blue green, complementary colors on the color wheel.

Are the dimensions of the King's Chamber intended to balance time and space by harmonizing frequency of time and measurement of space? Does this mean that we are in a zero time-space zone?

If we took the dimensions of a room in space and found the related frequency in time, would the illusion of the walls disappear? This is indeed what happened when the dimensions of a conference room were applied to the related frequency, and many of those present experienced that the physical walls disappeared.

LET US IMAGINE OURSELVES LIVING IN A SPACE SHIP

What does sound do in empty space? If there were truly a vacuum in empty space, then sound would not travel at all, as it would not even exist. Sound travels faster the denser the material it is traveling through. Through crystals, the speed of sound might be translated into an audible frequency of 272 cps, a musical note between P(C) and Q(D) that is close to a black note on the piano. The period of the rotation of the Earth around the sun is also calculated as 272 cps. (Reference: IEEE KIMAS 2005)

Now, let us suppose that five of us have boarded a space ship with little except the clothes on our backs and our instruments. We hear unfamiliar sounds that seem to be synchronized with different brilliant colors. We see webs of energy that seem to move with us all around our bodies. We wear seamless robes made of a fabric that registers every vibration from cell, to organ, to bone.

We choose our own favorite colors and our favorite keynote sound. Our spaces are constructed by our thoughts out of planes of light in different colors.

One plane of color forms a wall; it becomes yellow as we sing the note R(E). That is a color of the solar plexus that stimulates appetite. Another plane of color becomes violet as we sing the note W(B), a color and sound that is the crown chakra, an Etheric mental quality. We sing the note T(G) to create one more plane of color as blue. T(G) is the throat chakra, the energy of communication. We see these three planes of color forming an apex above us as we sing.

We sit in a multi colored tetrahedron, one of Plato's five solids, where the sounds of three chords are a triad. We see our triangular floor as red for the note P(C), a symbol for root chakra energy. Then, as we sing, the transparent walls of colored light begin to spin in unison, the illusion of the room becomes a cube, and we find ourselves in a space-time environment where we are singing our own energy music.

Once we create our own environment from our intent and our song, we visualize an object that we need, and sing it into being. We realize that every form is made up of different patterns of energy, and when we find its aural pattern it manifests as a shape.

We have just manifested a ripe pear, and we are drawing the energy of the pear into our own bodies. The pear is vibrating in its own harmonic pattern different from, yet unified with all other harmonic patterns of form.

Now it is time for us to explore the laws of harmonic and sub harmonic music. Since I define sub harmonic music as approaching the concept of gravity, with lower frequency pitches and longer wavelengths, how will we know how sub harmonic sounds will behave in a space vehicle where there is no gravity? The answer might be that we only hear harmonic music that ascends to the light in higher frequency pitches and infinitesimal wavelengths: in space, then, perhaps there is no duality, no sub harmonic, no gravity, only the harmonics of ascending frequencies that could lead to light and ascension to harmonic higher and higher consciousness and to spirit.

THE SELF-HEALING LAWS OF HARMONIC MUSIC ON EARTH AND IN SPACE USING THE "INSTRUMENT OF THE FUTURE"

Frequencies of self-healing harmonic sounds on Earth from the PLHK. Each of seven keynotes has its own frequency. The one that you like the most will be your own keynote, which may vary from day to day. Each keynote has its own color and its own shape.

What is the word that you speak when you hear your own keynote? What is the wavelength of your own keynote? The wavelength will depend on the pitch; if you are a female you will probably choose a higher pitch and a shorter wavelength than a male, and vice versa.

If you play seven octaves higher than your audible pitch, the wavelength will be smaller and smaller until they seem to disappear, and you may be in an etheric or spiritual realm of higher and higher consciousness leading to light. The lower in pitch that you play, the longer the wavelength will extend. If you play seven octaves lower than your audible pitch, the measurement will be longer and longer until it might lead to the magnetic force of gravity and matter.

SELF-HEALING VIBRATIONS

I think of music as the harmonic frequency of vibration according to an ascending number of smaller and smaller steps, or ratios, from 1/16 to the infinitely small, or as sub harmonic vibration frequency descending in lower and longer steps from 1/16 to the infinitely large in wavelength.

We think about the possibility of having only seven music notations with subtle pitch gradations that need to be chosen in one audible octave. We think of the possibility of having audible and sub audible octaves and an infinite number of octaves that are unheard, but nevertheless exist. We think of each octave as a multiple of 2, with a ratio of 2/1 or 1/2. We remember that in our Western music there are only seven notations with subtle gradations from high to low pitches within a chosen octave that may be chosen as keynote signatures. We play an unlimited number of pitches that are either several cycles per second higher or lower than the seven fixed intervals in music, until we find the resonant pitch, vibration, or frequency that enables us to heal ourselves and others.

The self-healing nature of harmonic interval sounds, according to observations from hundreds of individuals, comes as an individual plays harmonic interval chords that are resonant with their keynote choice on the PLHK. The individual players' keynote choice seems to be the one that resonates to an individual's emotional, spiritual, physical or oracle[2] needs, and the specific choice of frequency generates the entire harmonic field of 256 intervals in the matrix. This generation from a single frequency includes a range of frequencies and octaves from 16 cycles per second to more than 20,000 cycles per second, and resonates with cells, organs, and subtle energies within their bodies.

These seven gradations are frequencies that form healing chords based upon the harmonic laws of musical intervals. We play creating melody. We play together, rhythmically creating healing time signatures in ratios of beats per second.

RESONANT FREQUENCIES

I have also explored resonance frequencies based upon certain dimensions of time. For example, planets are in a resonance relationship to the sun. The period of the sun as it spirals around our galaxy takes 25,000 years to arrive approximately at the same place, perhaps an octave higher.

When that frequency is translated into velocity in cycles per second, the musical keynote frequency becomes 333 cps. E (R), similar to the frequencies of each planet, as calculated by second, minute, hour, day, year etc. by the physicist Hermann Helmholtz, in his definitive book, "On the Sensations of Tone." (Reference IEEE KIMAS 2005)

My hypothesis is that the resonant frequency of hydrogen, with only 1 electron in its orbit around its nucleus, is 256 cps. This puts hydrogen in an audible middle C (P) octave. Therefore, I deduce that all the other elements in the periodic table relate harmonically to this keynote. This hypothesis provides me with a simple way to calculate the frequencies of each element in the periodic table.

---

[2] The term 'oracle body' refers to the part of ourselves that self-diagnoses.

INSPIRING SELF-HEALING BY HEARING THE VIBRATIONS OF NUTRITIVE ELEMENTS FROM THE PERIODIC TABLE USING THE "INSTRUMENT FROM THE FUTURE"

I believe that just by listening to the vibration associated with mineral supplements such as iron, zinc, calcium, manganese, magnesium, and selenium, that these minerals might absorb into our cells.

I have put a set of nutritive elements on my website under "music." They were all recorded on the PLHK. I listen to these nearly every day, and notice that my appetite is decreasing for foods, and that I have lost weight.

The vibrations of these minerals might aid in the restructuring of the appetite, so that the vibrations of the minerals might absorb the vibrations of these minerals in food. The nutritive elements that humans need are transformed into self-healing sound vibrations.

Using the PLHK the intervals are all harmonically related to each other so that they inspire self-healing. This might not be true with other instruments. I have given PLHK workshops with a medical doctor, who has tested individuals "radionically" before they played the PLHK and after. She detected toxins before playing, but found that they were not present afterwards.

Now, let us suppose that each individual has his or her own vibrating keynote, each connected to their vibrating glandular energy centers. And let us also suppose that these vibrating energy centers are connected to the planets in our solar system. By drawing a picture of a human and comparing the positions of their energy centers (chakras) in frequencies and musical notations, with the frequencies and positions of the planets on the human body, we find a close similarity in frequency of energy to planets.

We are able to do this by using just one vibrating audible octave as a reference. The chakra system follows a system of root, polarity, solar plexus, heart, throat, third eye, and crown. Each chakra might be connected to an endocrine glandular system, and the spine follows another system with the glands and organs. These systems may all be inter-related by vibrational frequencies that correspond with musical harmonic notations, and even the orbits and spins of planets.

Eight New Chakras

Transpersonal 273 cps (16:15) C# (Earth's Orbit 272 cps)

Third Eye 448 cps
(14:8) Bb

Crown 480 cps
(15:8) B

Venusian (Orbit 442 cps)
448 cps (14:8) Bb

Venusian (Venus spin 410 cps)
410 cps (16:10) Ab

Psychic Center (Uranus Orbit 415 cps)
415 cps (13:8) Ab

Throat 384 cps
(12:8) I G

Gaia (Earth spin 378 cps)
372 cps (16:11) F#

Thymus 352 cps
(11:8) Gb

Heart 341 cps
(16:12) F

Solar Plexus 320 cps
(10:8) Eb

Diaphragm 315 cps
(16:13) D#

Saturnian (Saturn orbit 296 cps)
293 cps (16:14) D+

Polarity (Mars orbit 289 cps)
288 cps (9:8) D

Root 256 cps
(1:1) C

© Copyright 1996
by Barbara Hero

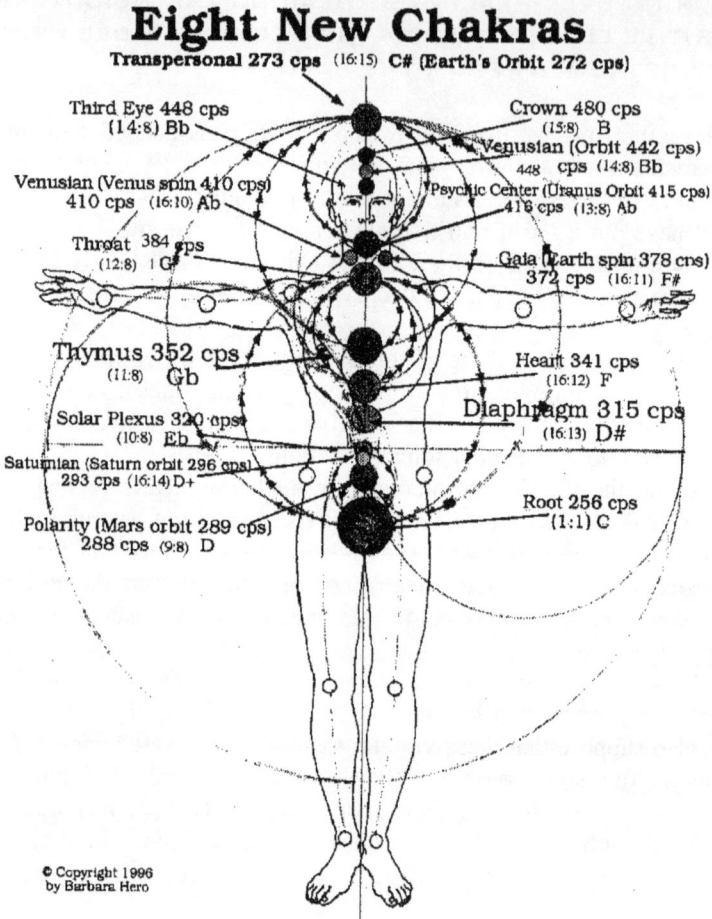

Figure 5
Chakras and Planets

This leads us to ask if there is a frequency that would cover all healing, and the answer seems to be that a healing frequency depends upon each individual's vibratory energy needs. These needs are assessed by the individual's choice of keynote; each individual makes his or her own choice.

Hence, a man once said to me, "What if I chose a keynote that I dislike?"

I responded, "Pick one that you dislike."

He did, and when I asked him to say a word that expressed that chord, he replied, "Death."

Most individuals select the keynote to which they like, resonate to, and need. The corresponding words that they choose include "I am," "clear," "angels," "mystery," "relaxation," "wonder," "joy, "peace," "excited," and "flying."

EFFECTS OF SELF-HEALING HARMONIC MUSIC ON EARTH AND IN SPACE

In order to be self-healing, PLHK music has a generating keynote that transforms a whole matrix field of eight musical notations from a pattern of an overtone and undertone series that relates to an individual's fundamental keynote.

The transformation happens upon finding the resonance between the body and the note. This may happen by humming along with music from the PLHK or from pure harmonic software based upon the Lambdoma matrix.

Recently a woman phoned me to say how she had been healed from a whiplash that had affected her neck and shoulders. She was listening to one of my self-healing sounds cassettes when she began to hum, and felt each chakra in her body resonating to each interval that she heard.

Think of the vibrating generating keynote as any one of seven notations that include bones in the spine separated with a left and right side of the body compared to the right and left side of the human brain. Think of the matrix field as made up of a fixed pattern of ratios where each ratio is multiplied by the vibrating chosen keynote. It is this transformation of a field by a single keynote that creates the healing energy potential of the matrix.

Each cell, each organ, each emotion, each etheric, each mental component, is influenced by the harmony of the whole matrix of ratios multiplied by its keynote, which mimics the patterning of our human body. This happens especially when an individual chooses his or her own keynote according to the pleasantness and pitch of the sounds to their own ears. This elicits a feeling of joyousness and awareness, and many remember their childhoods, or past lives and perhaps even future lives. For some, depression or anger has turned to relaxation and calm.

The music from this harmonic and sub harmonic pattern of ratios not only affects the emotional, spiritual and oracle bodies, but also often relieves physical pain.

Children who have experimented with these harmonic vibrations often compose their own melodies, while others reply to questions with the apparent wisdom of an elder.

In space, this keyboard could provide soothing experiences for relaxation, to solve problems, to listen to mineral supplements such as iron, zinc, calcium, manganese, magnesium, or selenium.

Games can be played, such as learning the shapes of individual harmonic intervals. When harmonic chords are played, the Lissajous shapes of the sounds become beautiful geometric figures that change with every chord. Some people are initially more interested in the shapes of the sounds than the sounds themselves, but when the shapes take on simple and beautiful patterns, they become aware of the relationship between harmonic sounds and beautiful shapes, and fix their attention on these healing vibrations that raise their consciousness to new levels of understanding of self and others.

CONCLUSION

The musical instrument that was brought to me from the future, the Pythagorean Lambdoma Harmonic Keyboard, suggests to me that perhaps there is not so much difference between the past, the present, or the future.

The advantage of living in space is that we can leave behind much of the baggage that we do not need or use on all levels of our being. The PLHK helps us to live in the moment, as we play it and watch the wavelengths of the chords and melodies change from one beautiful shape to another.

It becomes a kinesthetic experience that can induce a pleasantly altered state, one where there is no anger, only love. I have observed that for many people with problems to solve on the emotional, physical, spiritual or mental levels, solutions often come instantaneously. The instrument leads to a peaceful, harmonic way of life, as envisioned by Pythagoras in 500 BC, when he strummed his lyre and dissipated anger in one of his disciples.

We all need to know who we really are, and to live harmoniously among our fellow sentient beings. If art, music and mathematics can be combined to help us in our quests, let us take these tools on board for our adventures in space.

•••

BARBARA HERO

Founder/Director International Lambdoma Research Institute since 1997, Barbara is a visual artist, participating in exhibitions from the 1950's into the 1990's. Barbara is a composer and music theorist, having studied at the New England Conservatory of Music for eight years during the 1970s. Barbara received her Master's Degree in Mathematics Education from Boston University in 1981.

Barbara has combined art, music and mathematics in her research and in the invention of the "Pythagorean Lambdoma Harmonic Keyboard." She has written peer reviewed articles for the IEEE on the psychological benefits of the Lambdoma Keyboard from 1999 to 2007. From 1975 to 2007 Barbara has written for Art and Technology, Mathematics, Pavlovian, and Music journals, as well as chapters in several books on the philosophical mysteries of the Lambdoma matrix. Barbara gives lectures and workshops inspiring the self-healing effects of the Lambdoma Keyboard on individuals each year in the United States and several abroad from time to time. You may reach Barbara through her website, www.lambdoma.com.

REFERENCES

1. Hero B. Drawings Based on laser Lissajous Figures and the Lambdoma Diagram, *Leonardo*, Vol 11 pp. 301 -303, Pergamon Press Ltd. England, 1978.
2. Hero, B. F & Foulkrod R. M., Integrative music of the Lambdoma. *Integrative Physiological and Behavioral Science, The Official Journal of the Pavlovian Society*, 35(3), pp. 224-232, 2000.
3. Hero, B.F., R.M. Foulkrod: "The Lambdoma matrix and harmonic intervals, the physiological and psychological effects on human adaptation from combining math and music," *IEEE Engineering in Medicine and Biology Magazine*, Vol. 18 Number 2 March/April, pp. 61 – 73, 1999.
4. Hero, B: "The Pythagorean System: The Tetractys, the X and the Lambdoma," *Ambiguity and Music*, Editor, Jan haluska, Seminar Mathematics and Music, Bratislava, pp. 19 – 38, 1999.
5. Hero, B., "Eight Drawings of Tone Systems with Text." *Harmonic Analysis and Tone Systems*, Editor Jan Haluska, Tatra Mountains Mathematical Publications, Vol. 23, 2001 Mathematical Institute, Slovak Academy of Sciences, Bratislava.
6. Hero B. F. & Martinez E., "The Historical Context of the Lambdoma matrix and its Applications," KIMAS'05: Modeling, Exploration, and Engineering, Sponsored by IEEE Boston Section, in Cooperation with: *IEEE Systems, Man and Cybernetics Society*, INNs, US Air Force, US Army, US Navy, and DARPA, Editors Craig Thompson and Henry Hexmoor, pp. 411 – 417, 2005.
7. Hero, Barbara Ferrell, "Multidimensional Analysis of the Lambdoma Keyboard Experiments," 2007   International Conference on Integration of Knowledge Intensive Multi-Agent Systems, KIMAS'07: Modeling,

EVOLUTION and Engineering, Sponsored by: IEEE Boston Section in Cooperation with IEEE systems, Man and Cybernetics Society, INNS, US Air Force, US Army, US Navy, and DARPA, Editors: Henry Hexmoor and Craig Thompson, 2007.

CHAPTER 7

INVITATION TO ASTROSOCIOLOGY:
A NEW ALTERNATIVE FOR THE SOCIAL SCIENTIST-SPACE ENTHUSIAST[1]

© Jim Pass, Ph.D., 2009, All Rights Reserved

INTRODUCTION:
HOW DOES THIS INVITATION RELATE TO ME?

Space engineers and astronomers, as examples, may wonder how this chapter relates to their working lives. After all, their occupations do not seem to require knowledge about human behavior, or social patterns, or culture. Besides, the space age has progressed just fine for fifty years without a great level of contribution from social scientists. We even reached the moon in 1969 without much sociological or anthropological research.

This line of reasoning inevitably results in a straightforward question: How does this invitation to astrosociology relate to me? Similarly, why

---

[1] This chapter was adapted from a paper entitled *Invitation to Astrosociology: Why the Sociologist-Space Enthusiast Should Consider It* that the author presented at the 2005 ASA conference in Philadelphia as part of the Science, Knowledge, and Technology (SKAT) roundtables. The original paper is currently available at the *ARI* website at the following URL: http://www.astrosociology.org/vlibrary.html (in the "Astrosociological References" section).

should I care if the number of astrosociologists increases, or if the field even survives at all?

The reasons for interest by the space community are many. The following incomplete list, in random order, provides a good idea of just how wide-ranging and relevant astrosociology is to space exploration and related issues, especially for humanity's future expansion beyond Earth. Times are changing.

COLLABORATION

People are going into space, people live in social groups and must interact with one another, and therefore the social sciences are important (Pass 2005). Social scientists will contribute to the work of traditional members of the space community in ways that enhance progress that may be otherwise impossible to achieve. All of the benefits described below rely on effective collaboration.

THE UNKNOWN

What insights remain missing without information from the social sciences? We do not know what we do not know. Thus, a more comprehensive knowledge base – one that humanity will need to live in space – will emerge through the combination of both branches of science, the physical/natural sciences and the social sciences.

IMPACT OF SPACE SCIENCES

The untested ideas and scientific findings of philosophers and scientists have caused human cultures to rethink their self-images on a continual basis and to consider how humanity relates to the cosmos. It makes little sense to believe that this connection is less relevant today. Through an overt awareness of this connection, space scientists can better communicate the rationale of their work to the public, while social scientists can study its impact on society.

SOCIAL LIFE IN SPACE

Group dynamics in enclosed environments have already revealed "behavioral anomalies" to the space community. Solutions to mitigate negative social, cultural, psychological, and social psychological problems will require input from social scientists. The social environment requires recognition as the complement to the physical environment that served as the focus for engineers almost exclusively during the space age (Pass 2006b).

SPACE IN THE FUTURE

Who will go into space from this point forward? Will we send only robotic spacecraft? Will we send only professional personnel and elites who can afford to pay their passage? If a broader representation of humanity will go into space, then the same issues people face in terrestrial societies will reconstruct themselves in space environments in similar patterns.

MEDICAL ASTROSOCIOLOGY

Space medicine involves more than biomedical issues (Pass 2008). This new subfield considers such issues while also taking into account the social, cultural, ethical, and psychological aspects of space medicine. Medical issues possess broad ramifications for living on Earth and in space.

SPACE SOCIETIES

Space settlements are, in fact, space societies. Social life on Mars or elsewhere in the solar system will replicate many of the patterns of societies on Earth. Only social scientists possess the training and background necessary to provide the needed analysis to the space community.

TERRESTRIAL SPACEFARING SOCIETIES

Space and human activities associated with it have always affected societies and their cultures. As humanity expands into space, that impact will grow to a point at which, potentially, societies develop into spacefaring societies that feature close ties between space and social institutions, and a more centralized orientation to space by culture. Such social change requires study by social scientists in collaboration with space scientists (Pass and Harrison 2007).

PLANETARY DEFENSE

The focus of astrosociology expands beyond the traditional one that concentrates on the survival of Earth and the individual to include that of cultures, societies, and the human civilization. What steps are prudent to protect social life as we know it? Do we move people and infrastructure underground on Earth and on other cosmic bodies? Do we construct societies beyond Earth? A civilization-ending event could occur if we leave "all our eggs in the same basket." Preparedness for a large impact by an asteroid or comet rarely receives attention from an astrosociological perspective, yet we should begin serious discussions about it at the present so we can prepare ourselves.

SCIENTIFIC INTEREST

The possibility exists that a current member of the space community, whether studying in school or firmly entrenched in an aerospace company or space agency, may find social-scientific issues intriguing. Perhaps a change in focus occurs. Whatever the reason, even an engineer, architect, or space scientist may want to become an astrosociologist.

THE FOCUS IS SPACE

Historically, traditional thinking pits STEM subjects against everything else, including the social sciences. If one accepts any of the reasons included on this list, then one must bear in mind that the focus is space. Space exploration benefits when students focus on space issues, and because the two branches of science represent complementary approaches to space education and research, the overall effort must be on attracting students to study space issues from both orientations. When one considers this issue from this perspective, it is not a competition between the sciences, but rather a cooperative enterprise (Pass 2007).

In conclusion, then, traditional members of the space community have a real stake in the successful development of astrosociology. On the other side of the ledger, social science disciplines also have a stake in astrosociology. Historically, however, astrosociology – or a field similar to it – has remained the forsaken frontier for sociology and the other social sciences (Pass 2004c). As we move forward, the space community and the social science community both benefit by contributing to the successful development of astrosociology.

RELEVANT ILLUSTRATION: LIVING IN SPACE

The seventh reason above is especially relevant to this book's theme as it relates to settlements/colonies, or what I like to call "space societies," because of how human populations organize themselves on Earth. Sending a group of human beings into space to live creates a host of issues and problems with which humans continue to wrestle on Earth. The social sciences developed on terra firma long before the space age due to the need to understand human behavior in the context of larger social structures, and life in space will certainly demand the same.

Additional stress factors produced in space environments will amplify the need to understand social-scientific phenomena. For example, the fact that space settlers will find it impossible simply to breathe the "air" without assistance from technology will make social life more difficult. Considering the isolation, closed environment, unfamiliar physical location, different length of day and night, limitations of supplies, potential fragility of the life support system, reduced level of medical care, and other potential hardships, it is quite clear that establishing social life in space environments is more complicated than settling terrestrial lands.

Thus, *living in space* will require much more than the construction of a functional physical habitat. Additionally, when people are involved, complications develop beyond the requirements needed to conduct successful robotic missions.

If we truly intend to develop a space [society], we should remember one fundamental rule: *construction of the social environment is just as important for survival as construction of the physical environment.* The social construction of a space colony refers to the idea that settlements in space involve the creation of a social environment in addition to the physical environment (Pass 2006b:2). [italics from the original]

The development of a true space society requires a shared culture and social institutions to carry out important functions of social life. Within the larger structure of the society, communities will prove just as important beyond the Earth as upon its surface. Thus, a society is much more complex than a collection of people, and thus requires study by social scientists.

Psychologically, many individuals will find it difficult to fit into a space-born culture removed from terrestrial societies. The need to fit in and cope with the potential hardships of life in a newly constructed habitat would probably receive too little attention without input from social scientists. Even the design of the habitat, both externally and internally, involves many often-overlooked implications that relate to human behavior. The members of the population must function properly psychologically just as the habitat must function properly mechanically.

This invitation to astrosociology is especially pertinent to human space settlements of all types. Those seeking a permanent living arrangement beyond Earth will benefit enormously from the collaboration between astrosociologists and members of the space community, as the physical and social environments really involve a single system. Survival of the habitat means little if the survival of the society within it fails.

Living in space will likely not be possible without astrosociology, just as it is not possible without proper engineering or space architecture. Therefore, collaboration between both the aerospace and astrosociological communities represents the best chance at a livable space society. A high standard of living certainly must become the goal, rather than succumbing to an unbearable, dangerous subsistence.

PECULIAR PATTERNS OF INDIFFERENCE AND RESISTANCE

Sociology and the mainstream ranks of the other social and behavioral sciences and humanities – henceforth abbreviated as the "social sciences"– have largely refused to address astrosocial phenomena, i.e., social/cultural/behavioral patterns related to "outer" space.[2] A large element of this self-imposed indifference/resistance relates to the general assumption that this neglected area of social life is not legitimate and thus

---

[2] The "behavioral" component of the definition of *astrosocial phenomena* represents a recent update that acknowledges the growing participation of scientists and students in psychology and related fields who wish to participate in the development of astrosociology.

not worthy of sociological inquiry. The argument in this chapter takes the defiant stand that astrosociological issues should be considered legitimate, and moreover that any sociologist who is also a space enthusiast strongly interested in astrosocial phenomena must strongly consider astrosociology as the focus of his or her professional career (despite the potential negative consequences). This chapter invites sociologists and other interested scientists to join in breaking down the boundaries currently excluding astrosociological issues from the mainstream discourse within the sociological and other social science communities. It is time for the social sciences to re-examine its patterns of indifference and resistance.

For the development of astrosociology to move forward as a viable multidisciplinary field, the effort must include a most important continuing goal: the recruitment of students and other professionals to become formal astrosociologists, as well as encouraging decision makers running existing programs and departments to integrate this new field into their curricula. Astrosociology must enter academia.

Admittedly, the development of this new field faces huge challenges, though several signs exist that social scientists, especially social science students, view astrosociology as an exciting new subject area worth pursuing. This chapter serves as both an argument for the pursuit of astrosociology as well as an appeal to do so. Everyone interested in space should be aware of it.

As the founder of this new field, I have a compelling reason to promote it as if it represents a "no brainer" long overdue. Whether I can succeed in justifying its development and can spread the word beyond a rather small contemporary group of early supporters will largely determine the course of this new movement begun in July 2003. This invitation sets the stage for what comes next in astrosociology's development. Here, I provide reasons to pursue astrosociology that hopefully demonstrate the high value of interesting and relevant specializations open for pioneering work in a wide-open new field.

WHY STUDY OUTER SPACE ISSUES FROM A SOCIAL-SCIENTIFIC PERSPECTIVE?

The question often arises as to why humanity should explore space. The answers abound in the literature, and thus fall outside to the purview of this discussion. Suffice it to state that there are long lists of reasons why the human species will expand beyond the confines of its home planet – and why it should – that range from philosophical justifications, to practical necessities, to humanity-protection rationalizations. There are many more tantalizing reasons to explore space than reasons to isolate humanity from the rest of the universe.

Beyond that, space activities impact positively on societies on Earth through technology transfers and spinoffs, as well as in less tangible yet equally important ways. There is a good possibility that space exploration, settlement, exploitation, jobs, and recreation could over time transform patterns on Earth to resemble the characteristics of an ideal type of spacefaring society.[3] Space issues would then become paramount. This would then require the input of astrosociological research to guide space policy and increase the level of understanding about relevant developments.

This chapter has a twofold purpose, presented in reverse order of its title. First, it addresses a fundamental question: why should a space enthusiast consider becoming an astrosociologist in the current subcultural climate characterized by indifference, resistance, and even possible negative consequences for doing so? Second, while it serves as a general invitation, it also offers targeted invitations to space enthusiasts of various backgrounds. Those considering sociology or any social science are encouraged to become astrosociologists, including students, newly trained social scientists, and established professionals willing to change their areas of concentration.

Astrosociology must be seen as a legitimate and essential field before (1) a significant number of social scientists will consider changing their focus and "switch" to this new field, (2) departments within universities and colleges will consider adopting it, and (3) students will consider selecting astrosociology as an area of concentration (though there is already evidence for the growth of this third trend).

As such, this early invitation extends to those who are willing to accept a potentially difficult challenge. This open invitation intends to encourage well-regarded scientists who are also interested in "outer" space (as opposed to "social" space or any type traditionally studied by social scientists) to help forge the growing astrosociological community. The first few of those hearty souls who openly characterize themselves as astrosociologists must be willing to take a calculated risk. Those in the more powerful positions within their disciplines, as well as those at lower ranks but possessing high levels of prestige, take the greatest risk. At the same time, they can also potentially provide astrosociology with the greatest boost of legitimacy.

The notion of this invitation, a title modified from Peter Berger's (1963) *Invitation to Sociology*, is presented here in the same spirit and earnestness. This chapter does not pretend to make the same arguments as Berger, although the objective of applying a sociological (and broader social science) understanding of astrosocial phenomena does reflect a specific application of his general argument. Here the focus is on (1) the

---

[3] See Pass and Harrison (2007) for additional details.

failure of the sociological discipline to apply his invitation to human activities related to space (also reflected in the other social sciences) and (2) the need to address this problem. In other words, it amounts to reissuing Berger's invitation with direct application to astrosociology and all of the social and behavioral phenomena under its purview (Pass 2004a). The need to focus on astrosocial phenomena is just as urgent as his general appeal to join sociology.

A SOCIOLOGICAL BIAS...

My training in sociology places a filter on my perspective, though I believe my outlook is expanding as I work with scientists and scholars in other fields, and as astrosociology continues its development. I look forward to increasing my exposure to complementary perspectives that support this new field. Astrosociology ties together seemingly unrelated fields due to its focus on both space and humanity.

Rigorous sociological inquiry has largely failed to focus upon astrosociological issues on a consistent basis due to three related reasons. Sociologists perceive them as (1) lacking legitimacy, (2) existing on the "fringe" of social life, and (3) possessing few important consequences for society (Pass 2004c). This chapter assumes that astrosociology, although new to sociology (and more generally to the social sciences), is a legitimate area of sociological inquiry that does not differ from other well-established subfields that currently enjoy large followings. The space program of any space-capable society involves many institutions, larger culture and subcultures, and a great number of people. Thus, it possesses "strong roots" in that society making it worthy of serious investigation.

A BRIEF HISTORY OF ASTROSOCIOLOGY[4]

From the beginning of the effort to develop this new field, we noticed an undercurrent of people conducting research and otherwise interested in space issues from a social-scientific perspective. They existed in isolation and normally apart from their mainstream colleagues, but continued to work in their areas of interest. For fifty years, something was missing though.

Before uploading the original version of the website dedicated to the development of astrosociology, Astrosociology.com, on July 15, 2003, it took a few months of thought and discussion with others to define the

---

[4] For a complete history of astrosociology, see the Calendar/History page at Astrosociology.org (link: www.astrosociology.org/calendar_history.html).

parameters of astrosociology; including the boundaries that identify what lies within its scope and what lies outside of it. The concept of astrosocial phenomena, proved elusive for a short time, as did its definition, though it came to define the purview of astrosociology fairly well in the end. Many individuals already interact with the Astrosociology Research Institute (ARI)[5] via email or through its website Astrosociology.org (which took over for Astrosociology.com) as either "members" or "supporters." They are sociologists, other types of social scientists, and those from the space community.

Interestingly, space engineers, architects, and other non-social scientists seemed more interested in astrosociology than the mainstreams of the social science disciplines from the beginning, even after more than five years. Once exposed to it, the space community promptly supported the field of astrosociology. After briefly introducing the field to a Technical Committee of the American Institute of Aeronautics and Astronautics (AIAA), the Astrosociology Working Group (AWG) soon formed. This places astrosociology within the structure of the AIAA and thus in a good position to both attract new supporters and provide needed social science research findings to a professional organization that mostly consists of engineers, space scientists, and others oriented toward the natural and physical sciences, and who have little or no exposure to the social sciences.

We have presented papers and held sessions on astrosociology at conferences including the AIAA Space Conference, the AIAA Aerospace Sciences Meeting (ASM), and the Space Technology and Applications International Forum (STAIF). In February 2009, this author served as chair of the 1st Symposium on Astrosociology at the Space Propulsion and Energy Sciences International Forum (SPESIF). These developments demonstrate that members of the space community have begun to recognize that they must collaborate with social scientists in order for humanity to survive beyond the relatively safe confines of its home planet. Presentations at conferences oriented toward both branches of science reveal the potential of astrosociology to bridge both branches of science and allow for structured collaboration.

With many successful experiences already behind us, we move forward with confidence in the successful development of astrosociology. We realize that the road will not be an easy one, though we remain confident that (1) the void represented by the sparse coverage of astrosociological issues will become evident to others and (2) a substantial number of scientists will join this movement to remedy the situation. This invitation serves to inform members of the space and social science

---

[5] The Astrosociology Research Institute (ARI) is a California nonprofit public benefit corporation with a mission to conduct astrosociological research and assist others by various means to develop astrosociology as a viable scientific field within academia. The website URL for ARI is http://www.astrosociology.org.

communities of important new opportunities in a field vital to the success of long-term space exploration and settlement.

WHY SHOULD A SOCIAL SCIENTIST-SPACE ENTHUSIAST BECOME AN ASTROSOCIOLOGIST?

The simple answer is "because it makes logical sense." A strong interest in space exploration or space science should encourage a social scientist to pursue astrosociology as the focus of his or her vocation. This is as natural as the following: Why should a basketball fan become a sports sociologist? Why should the fascination with criminal behavior compel a sociologist to become a criminologist? Why should concern about the various types of medical ethics abuses cause one to become a medical sociologist?

Astrosociology is the scientific application of a social-scientific perspective to the objective analysis of astrosocial phenomena, and while a space enthusiast can become an astrosociologist, this is not to imply that all astrosociologists must be space enthusiasts. One can study astrosocial phenomena without the need to be obsessed with space.

Astrosociology must be a field of social science and not simply a forum for advocating greater funding for various space missions and programs. Astrosociology is a subfield of sociology and a multidisciplinary field, and it must therefore follow the same strong tradition of conducting science objectively.

Can a space enthusiast, who also happens to be an astrosociologist, produce truthful research findings when a space policy decision is objectively bad or a space project is over budget without any chance of meeting its objectives? More generally stated, will personally held values overwhelm (or at least taint) objective science? This is akin to asking whether a sociologist, who happens to be a Catholic, can conduct research about a cult without imposing his or her religious values onto that group, or attempting to affect its worldview. The sociology of religion is an important subfield in which sociologists from all backgrounds receive an open invitation to participate. Any good scientist, including an astrosociologist, would experience few problems in producing research findings that reflect the reality as depicted in the data.

Supporting a decision or program is not in the job description of a scientist; at least within the role set of scientist. This is the job of a space lobbyist. An astrosociologist is a scientist and thus goes where the data takes him or her, without succumbing to bias that could affect the outcome of the research effort.

The particular organizational environment in which an astrosociologist works should also make no difference. Applied astrosociology is likely to become a common approach. An astrosociologist can work on a SETI project, for example, and still be unbiased in the search for understanding of how SETI efforts influence society, constructing interstellar messages, studying the implications of a detected signal, contemplating the possible characteristics of an extraterrestrial civilization, or determining what the very search effort itself implies about the culture of a particular society. Gauging public support is an important function because it has implications about how an organization dedicated to SETI operates. Education efforts are also part of its normative functioning, and an astrosociologist may be involved in carrying out various surveys and ongoing program evaluation efforts in an objective manner.

PERSONAL INVITATIONS TO ASTROSOCIOLOGY

Formally accepting the status of "astrosociologist" currently represents a difficult proposition, although the newcomer will not be alone. A growing number of sociologists and scientists from the other social sciences, in addition to a growing number of professionals from the space community, indicate strong support for the establishment and successful development of astrosociology as a new field, even though most of them are not yet committed to formally referring to themselves as "astrosociologists." Overall, progress continues despite traditional barriers.

Does this call for the establishment of astrosociology represent a self-fulfilling prophecy or a fool's errand? No one can predict the future flawlessly. If astrosocial phenomena indeed reflect ignored yet significant social and behavioral patterns, then the self-fulfilling prophecy scenario correctly predicts the future course. However, interested scientists need to participate in order to ensure success because one only knows that a self-fulfilling prophecy has occurred following its aftermath!

Granted, this proposed multidisciplinary social science field is relatively new, so very few social scientists are yet aware of its existence, and it must therefore gain legitimacy to attract additional scientists. Until incorporated into the curricula of higher education organizations, for example, it will remain a difficult choice. But a space enthusiast who is also a social scientist must strongly consider astrosociology despite its legitimacy crisis, because it does exist and thus represents a real choice: Astrosociological pioneers remain indispensible to the future of the field.

My call is for others who are interested in this new field to join the dialog, to create an altered environment within the sociological subculture and those of the other disciplines. The only way to overcome this seemingly intractable situation is for a substantial number of others to join the cause of making astrosociology a well-known, well-respected multidisciplinary field, a long overdue social reality considering the potential future impact of astrosocial phenomena on society (Pass 2004b).

As the future unfolds, reliance on space exploration and space resources will likely increase due in part to their contributions to increasingly difficult social problems (Pass 2006a). The simple increase of astrosocial phenomena in terms of their ubiquity and significance speaks directly to the relevance of astrosociology despite its long standing absence.

A SPECIAL INVITATION TO INTERESTED SOCIAL SCIENCE STUDENTS AND SOCIAL SCIENTISTS

Renowned sociologist Peter Berger believed that sociologists should involve themselves in all possible levels of society based on a personal interest (though with an objective or value-neutral) approach, whether these social phenomena reflect personal values, or tastes, or even result in revulsion. One of the important lessons is that all aspects of social life are important, and should receive acceptance, because an investigator deems it so.

Thus his questions may lead him to all possible levels of society, the best and the least known places, the most respected and the most despised. And, if he is a good sociologist, he will find himself in all these places because his own questions have so taken possession of him that he has little choice but to seek for answers. (Berger, 1963:18).

In astrosociology we have a "level of society" that requires the same understanding as all others, and I invite all sociologists and other social scientists as well as those in the humanities who possess an interest in astrosocial phenomena to consider astrosociology.

Consider that no one can legitimately argue that astrosocial phenomena are not real social phenomena. These social patterns, like all others, are worthy of sociological inquiry because they exist (Pass 2004c). Purposeful avoidance of an area of investigation weakens any science and therefore deserves rectification. Astrosocial phenomena will only increase in importance to social life and thus we must investigate their implications.

To become an astrosociologist means to set out on a voyage of an unexplored social science territory; one that was always there, although never officially acknowledged within any discipline. One may apply the concept of final frontier to the undiscovered territory related to the social sciences as well as the cosmos (Pass 2004c), and consider this unexplored territory the astrosociological frontier, the still unfamiliar and little-explored territory characterized by astrosocial phenomena within the institution of higher education. The final frontier and astrosociological frontier require simultaneous exploration.

The creation of the Astrosociology Research Institute (ARI) in 2008 seeks to put interested parties on the correct path to remedy this situation by conducting astrosociological research, providing direct support to students and the other astrosociological pioneers, and seeking cooperation with colleges and universities to recognize this new field. ARI exists to provide resources and opportunities for astrosociological pioneers in their quest to assist in the development of their newly chosen field.

An easy way to introduce astrosociology into higher education can begin with students conducting research for class projects and assignments. I strongly encourage interested students to select a particular category of astrosocial phenomena as their topic, as perhaps it is a bit easier for students rather than established sociologists to do this because they have no reputation to protect. On the other hand, if students encounter professors who decline to accept astrosociological issues as legitimate topics for research papers, theses, or dissertations, they should still urge their professors to allow them to select topics falling under the purview of this new subfield. Students' actions may well develop into the driving force behind the successful development of astrosociology.

I invite interested faculty members to encourage their departments to adopt astrosociology, including introducing astrosociological issues into their courses, and encouraging interested students to follow their interests in the area. The best way to make headway into academia is through the participation of both students and faculty. This invitation requests that both join in concert in order to place pressure on their departments to accept change.

A SPECIAL INVITATION TO DISINTERESTED SOCIAL SCIENTISTS AND CRITICS

For those who remain indifferent because astrosocial phenomena do not interest them, I invite you to support the development of astrosociology in a general sense. When interested students and faculty request their desire to pursue astrosociology, please support the legitimacy of this field and its right to exist. Do not dissuade students or others from pursuing astrosociology merely because you have no interest in it.

Critics, to the extent that they provide rational negative analyses, can serve a positive function. You are welcome to continue questioning the need and legitimacy of astrosociology while doing nothing to inhibit its establishment or development for those who favor it. In this way, perhaps we can construct a better field that is sensitive to the concerns of critics. The goal in this respect remains to improve the social science disciplines for everyone involved.

A SPECIAL INVITATION TO THE DISCIPLINE/DEPARTMENTS

Social control within the subcultures of social science disciplines have played a part in keeping astrosociological issues from entering into the common discourse among sociologists and other social scientists. No social group is immune to politics. As such, longstanding structural social patterns within social science disciplines may contribute to the evaluation of astrosocial phenomena as illegitimate elements of social life. In this context, one must ask whether the "gatekeepers" in powerful positions within the various social science disciplines promote a negative view of astrosociology, even if unintentionally.

Sociologists and other social scientists should reconsider the value of astrosocial phenomena. If social science students reflect the rest of the population of the United States, then an appreciable majority of them are interested in space exploration generally and the space program particularly. By tapping into this significant level of interest, and even promoting it as an important component of their curriculum, departments can potentially increase the number of those taking sociology courses and even those majoring in sociology and the other social sciences.

At some point in the near future, assuming that astrosociology receives support from the social science community, departments will naturally consider creating courses dedicated to this new field. For example, as an initial step, I encourage departments to request their faculty to mention astrosociology when they lecture about the sociology of science and technology and when discussing new trends in the discipline. Students should be encouraged to write research papers, theses, and dissertations focusing of astrosocial phenomena.

A SPECIAL INVITATION TO INTERESTED NON-SOCIAL SCIENTISTS

Those interested in space in other capacities, such as engineers, astrobiologists, astronomers, would benefit by considering astrosociological issues in their everyday working lives. Futurists, for example, possess diverse backgrounds including the social sciences and space sciences. Many of these scientists express interest in social-scientific issues through their writings and verbal expressions, though without any direct exposure to the social sciences. Familiarity with astrosociological concepts and research findings can improve this discourse. Therefore, non-social scientists can benefit from astrosociological insights and thereby contribute to astrosociological knowledge.

As Berger points out,

> Sociological understanding can be recommended to social workers,
> but also to salesmen, nurses, evangelists and politicians – in fact, to
> anyone whose goals involve the manipulation of men, for whatever
> purpose and with whatever moral justification. (Berger 1963:5).

Using C. Wright Mills' (1959) concept of the sociological
imagination as an additional element of this argument, one can suggest that
everyone can benefit from a sociological understanding of a particular set
of social phenomena. With some exposure to the social sciences, non-
sociologists – whether possessing social science backgrounds, physical
science backgrounds, or any other – can contribute specifically to the
advancement of Astrosociology.

In summary, then, an invitation goes out to those of you willing to
incorporate relevant social-scientific principles into your work in order to
collaborate knowledgably with social science-trained astrosociologists in a
multidisciplinary, cooperative approach to understanding astrosocial
phenomena from a combination of related perspectives. A strictly
sociological perspective fails to represent the most practical, or most
productive, approach to constructing this new field. As interested non-
sociologists who have much to contribute, your participation serves as a
great additional dimension to the overall understanding of astrosociological
issues.

CONCLUSION

Berger points out that as social beings, we find ourselves controlled much
like puppets on strings, though with a decidedly different potential than
puppets.

> We locate ourselves in society and thus recognize our own position
> as we hang from its subtle strings. For a moment we see ourselves as
> puppets indeed. But then we grasp a decisive difference between the
> puppet theater and our own drama. Unlike the puppets, we have the
> possibility of stopping in our movements, looking up and perceiving the
> machinery by which we have been moved. In this act lies the first step
> towards freedom. And in this same act we find the conclusive justification
> of sociology as a humanistic discipline. (Berger 1963:176).

We can expand this observation to point out an important reason to
pursue the study of astrosocial phenomena: because humans can understand
their own manipulation, they can understand their mistakes and correct
them. They need not continue the error of their ways based simply on

historical inertia. Hence, adoption of astrosociology is possible, and even necessary, for the ultimate success of human space exploration.[6]

Based on the foregoing discussion, a final question requires consideration. Why would a space enthusiast who is also a social scientist (or student) refuse to consider seriously the pursuit of astrosociology? The answer lies in its lack of legitimacy, and thus the very practicality of it, as discussed.

But all unfamiliar forms of social-scientific inquiry often begin under difficult circumstances. Taking the first groundbreaking steps reflects a most difficult commitment due to the absence of an academic safety net. The early astrosociological pioneers must lay down the foundation for a well-respected field so that others can follow at some future point without the fear of harming their careers or worrying that astrosociology will never achieve acceptance within the social science disciplines. As the early astrosociological pioneers, they will be the ones who receive the greatest criticism based on their overt attempts to demonstrate the long-denied legitimacy of studying astrosocial phenomena.

If you meet the criteria of an astrosociological pioneer, please consider this an invitation to participate in astrosociology and help nurture its potential as an important new subfield. If you are not a social scientist through education or experience, but willing to incorporate a sociological or other social science perspective that focuses on astrosocial phenomena, your invitation to join the astrosociological community remains open as well.

In this appeal, I call for the reinterpretation of Mill's (1959) historic call for social reform as the embodiment of social science reform – that is, the restructuring of the social science disciplines themselves through strong advocacy aimed at adopting astrosociology as a mainstream, well-regarded field.

I extend a hearty welcome to those who wish to participate in the construction of this new and exciting field. For those who do not accept this invitation, due to a lack of interest in space, I only request that you consider supporting astrosociology's right to exist as a legitimate field in a general way; that is, allowing it to prosper within social science and even non-social science departments and programs without your interference. Astrosociology is a field that can stand on its own merits, and it should therefore receive serious consideration for the full support of the social science community and adoption to some extent by other disciplines as a contributing perspective to their substantive areas. The success of its introduction into higher education will benefit the social science community in countless ways, and we must not forget that it will also

---

[6] For example, an attempt to establish *exo-sociology* as a new section within American Sociological Association (ASA) failed in the past (Pass 2004b).

provide new theoretical and practical knowledge for the space community as well. Living in space successfully will require the best from scientists in both branches of science working together towards a common goal. The human dimension must receive its due attention in the study and planning of human space exploration and settlement. The time has come to expand our approach to understanding the evolving relationship between humanity and outer space.

•••

JIM PASS, PH.D.

Jim Pass received his doctorate in sociology at the University of Southern California in 1991. As a long-time avid space enthusiast, he always longed to combine his interest in space exploration with sociology. In December 2002, he came across an online article by Allen Tough about the social implications of SETI that mentioned the importance of establishing a new field that focused upon the social ramifications of space exploration. Dr. Tough mentioned astrosociology as one of two possible names for such a new field. Upon seeing this term, Dr. Pass purchased the domain of Astrosociology.com and set out to define this new field. Several months later, on July 15, 2003, he uploaded the website pages for Astrosociology.com. Development of astrosociology continues within the sociological, general social science, and space communities.

Dr. Pass has taught sociology at Long Beach State and Long Beach City College in California for more than fifteen years. Today, Dr. Pass devotes most of his time in advancing the development of astrosociology. For example, he was in great measure responsible for establishing an AIAA Astrosociology Working Group and serves as the Vice Chair. In May 2008, he – along with Dr. Marilyn Dudley-Flores and Thomas Gangale – established the Astrosociology Research Institute (ARI) as a nonprofit public benefit educational corporation with the mission to develop astrosociology and to assist others, including students, in doing so.

REFERENCES

1. Berger, Peter L. (1963). *Invitation to Sociology: A Humanist Perspective*. New York: Anchor Books, Doubleday.
2. Dudley-Rowley, Marilyn (2004). The Great Divide: Sociology and Aerospace. (Posted at Astrosociology.org). (Presented on October 16, 2004 as part of a special dedicated session, entitled "Astrosociology: The Establishment of a New Subfield," at the California Sociological Association (CSA) conference in Riverside, CA). URL: http://www.astrosociology.org/Library/ PDF/submissions/The%20Great%20Divide_CSA2004.pdf
3. Gangale, Thomas E. (2004). Practical Problems in Astrosociology. (Originally posted at OPS-Alaska.com and Astrosociology.org. (Presented on October 16, 2004 as part of a special dedicated session, entitled "Astrosociology: The Establishment of a New Subfield," at the California Sociological Association (CSA) conference in Riverside, CA). URL: http://www.astrosociology.org/Library/PDF/submissions/Practical%20Proble ms%20in%20Astrosociology.pdf
4. Harrison, Albert A. (2005). Overcoming the Image of Little Green Men: Astrosociology and SETI. (Paper presented on November 11, 2005 as part of a dedicated session, entitled "Astrosociology: The Sociology of Outer Space," at the California Sociological Association (CSA) conference in Sacramento, CA). URL: http://www.astrosociology.org/Library/PDF/submissions/Overcoming%20LG M_Harrison.pdf
5. Mills, C. Wright (1959). *The Sociological Imagination*. New York: Oxford University Press.
6. Pass, Jim (2008). Space Medicine: Medical Astrosociology in the Sickbay. (Paper presented on January 08, 2008 as part of a session entitled "Astrosociological Perspectives on Space Exploration" at the AIAA Aerospace Sciences Meeting and Exhibit (ASM 2008) conference in Reno, NV. It was also published in the proceedings). (URL: http://www.astrosociology.org/Library/PDF/ASM2008_MedicalAstrosociolog y.pdf
7. Pass, Jim (2007). Enhancing Space Exploration by Adding Astrosociology to the STEM Model. [Paper presented on September 20, 2007 as part of a session entitled "Astrosociology II: Space and Society" at the AIAA Space 2007 conference in Long Beach, CA. It was also published in the proceedings]. URL: http://www.astrosociology.org/Library/PDF/Contributions/Space%202007%2 0Articles/Adding%20Astrosociology%20to%20STEM%20Model.pdf
8. Pass, Jim (2006a). The Potential of Sociology in the Space Age: Developing Astrosociology to Fill an Extraordinary Void. [Paper accepted for presentation at Pacific Sociological Association (PSA) conference in Los Angeles, CA as part of the Sociology of Science and Knowledge session). URL: http://www.astrosociology.org/Library/PDF/submissions/Potential%20of%20 Astrosociology.pdf

9. Pass, Jim (2006b). "The Astrosociology of Space Colonies: Or the Social Construction of Societies in Space." Space Technology and Applications International Forum (STAIF) Conference Proceedings, Volume 813, Issue 1, pp. 1153-1161. (Paper published in Proceedings and presentation at 2006 STAIF Conference in Albuquerque, NM). URL: http://www.astrosociology.org/Library/PDF/submissions/STAIF_Astrosociology%20of%20Space%20ColoniesPDF.pdf

10. Pass, Jim (2005). "Astrosociology and the Space Community: Forging Collaboration for Better Understanding and Planning." The Space Review, Monday, August 8, 2005. URL: http://www.thespacereview.com/article/424/1

11. Pass, Jim (2004a). Inaugural Essay: The Definition and Relevance of Astrosociology in the Twenty-First Century (Part One: Definition, Theory and Scope. (Presented at Informal Roundtable at the 2004 ASA conference in San Francisco, CA). URL: http://www.astrosociology.org/Library/Iessay/iessay_p1.pdf

12. Pass, Jim (2004b). Inaugural Essay: The Definition and Relevance of Astrosociology in the Twenty-First Century (Part Two: Relevance of Astrosociology). (Presented as the lead paper as part of a special dedicated session, entitled "Astrosociology: The Establishment of a New Subfield," at the California Sociological Association (CSA) conference in Riverside, CA). URL: http://www.astrosociology.org/Library/Iessay/iessay_p2.pdf

13. Pass, Jim (2004c). Space: Sociology's Forsaken Frontier. (Presented at the 2004 California Sociological Association (CSA) conference in Riverside, CA). URL: http://www.astrosociology.org/Library/PDF/submissions/Space_Sociology%27s%20Forsaken%20Frontier.pdf

14. Pass, Jim, and Harrison Albert A. (2007). Shifting from Airports to Spaceports: An Astrosociological Model of Social Change toward Spacefaring Societies. [Paper presented on September 20, 2007 as part of a session entitled "Astrosociology I: Theory and Research" at the AIAA Space 2007 conference in Long Beach, CA. It was also published in the proceedings]. URL: http://www.astrosociology.org/Library/PDF/Contributions/Space%202007%20Articles/Airports%20to%20Spaceports.pdf

CHAPTER 8

INTERGENERATIONAL PATHWAYS
GETTING ALONG IN SPACE TAKES ON NEW MEANING

CATHY L. WONG, PH.D.

The frontier of space has opened up many possibilities for the development of new communities, and one of the specific areas that I find particularly interesting about these communities is the relationships between older and younger generations, and particularly the possibility of intergenerational collaborations. I believe that such collaborations will be essential to our success in establishing vibrant communities as we begin to think about working, living, and someday raising families in outer space.

In this paper I will explore the possibilities of forging what I have termed "intergenerational pathways" as a way of uniting older and younger generations to enhance the sharing of skills, knowledge, and innovations; thereby providing necessary social support to people in outer space, and on Earth as well.

Or, to say it another way, how might our society create, encourage, and prepare a new generation of thinkers and explorers who seek to colonize the frontiers of outer space?

For surely the right preparation will be essential to their success, and will open doors for those who have not yet tapped into their own latent talents. Ideally, this would lend itself to globalized support networks that will ensure an on-going space program from generation to generation.

REASONS TO CONSIDER COMING TOGETHER

As an aging Baby Boomer and sociologist, I am quite aware of the increasing older population that is already having a significant impact on all aspects our social life. U.S. Census projections have estimated that there will be 1.2 billion people age 65 years and older by 2025 (Atchley and Baruch, 2004 p. 31), so the impact will only increase. In conjunction, as new medical and scientific advances make it possible for people to live longer than ever before, their attitudes are changing, as is their role in society. Older adults are showing increasing interest in generative activities that "give back to society" to benefit themselves and others (Moraitou, Kolovou, Papasonzomenou, and Paschoula, 2006; Atchley and Baruch, 2004; Keyes, 1998). This reflects a growing sense of responsibility, and a desire to shift away from a focus on the individual, and toward a more inclusive collectivity.

As the challenges we face grow ever more demanding, no generation stands alone. All living generations must work together to create a culture that will allow future generations to flourish, and perhaps the key to working together, both on Earth as well as in outer space, is to embrace the notion that we are all in this universe together, rather than as separate human beings, separate generations, separate enterprises, or separate countries.

One way that I propose to create pathways to bring young and old together is through "learning that transforms," which has also been referred to as "transformative learning." First coined by Jack Mezirow, transformative learning "involves experiencing a deep, structural shift in the basic premises of thought, feelings, and actions."

On a micro-level, engaging in transformative learning might have a dramatic influence in how we view ourselves within the world, as we begin to see that we effect and are affected by the way we conduct our lives on a daily basis. On a macro-level, we might begin to understand the broader patterns of society, the "interlocking structures of class, race, and gender" (Morrell and O'Connor, 2002 p. xvii), as well as the value of linkages between young and old.

REAL AND PERCEIVED SEPARATIONS BETWEEN THE YOUNG AND THE OLD

People in most societies engage in some sort of division of labor, where segregation occurs due to age, and distinct rights and responsibilities are recognized at each stage (Hagestagd and Uhlenberg, 2006). Sociologists often refer to this as age stratification, "the division of people into categories or strata according to age—'young,' 'middle-aged,' and 'old'" (Hooyman and Kiyak, 2005 p. 291). This is particularly evident in tribal communities where the division of labor often separates men, women, children, and elders in distinct tasks. Men are often hunters, women gatherers, older children help out with the care of younger ones, and the oldest members take care of babies and engage in light chores around the encampment.

Age stratification in modern societies, however, does not generally lead to social integration, but rather to separation, isolation, and in many cases to prejudices and discrimination; which is often referred to as ageism. Hagestad and Uhlenberg's (2006) research on age and social integration exposes compelling reasons to take age segregation seriously; showing that "segregation affects individuals who are in quite different phases of life, while ignoring or making invisible those who might not fit any specific categorical role" (p. 643).

Hence, when a society legitimizes separation and segregation between groups, it can have adverse affects upon the lives of individuals. Hagestad and Uhlenberg (2006) outline these and the impact they can have between the young and old.

It produces and reproduces ageism

Ageism is a prejudice (pre-judgment) that is generally negative. It is supported by stereotypes that suggest that older people are "valueless, poor, frail, burdensome, asexual, and powerless or domineering" (Markson, 2003 p. 166). And while it is most common to discuss ageism in reference to the older population, younger generations also can become stigmatized in this manner. This is apparent when older people suggest that "the youth of today are spoiled," "lack motivation," or are "lazy." Prejudice leads to discriminatory acts upon both older and younger populations; thus alienating, oppressing, or simply ignoring the contributions that both generations can make to society, and legitimizing the separations between generations.

Separation is a threat to embeddedness, to the integration of all groups into society, and, for the old, it increases the risk of isolation.

The faulty idea that a person is only productive when he or she is engaged in paid work is deeply rooted in the American value and belief system. Thus, embeddedness is potentially reduced when someone's identity is directly linked to a value and belief system that rewards only the production of employees; while devaluing contributions made through volunteer, charity, parenting, grandparenting, or other nonpaid productive activities during the later years of life. One adverse affect upon older populations may be that they may turn inward; leading to isolation from others.

IT THWARTS SOCIALIZATION FOR YOUNG AND OLD

Throughout history, older people often played significant roles in the social lives of the young, and by acting as sages and seers, leaders, judges, and guardians of traditions, elders were often regarded as the transmitters of knowledge to younger generations. This, however, changed with the industrial revolution. As society's emphasis on technology grew, so did the gap between older and younger generations; as young people mastered technologies that their elders had no experience with. This intergenerational gap is apparent in the ways that both young and old view each other's place within society. Thus, it contributes to the tensions between generations. Negative myths and generalizations may also contribute to ageism.

IT IMPEDES GENERATIVITY

Separation may inhibit the concern of older generations for the young, also know as generativity, and therefore it also impedes the creation and maintenance of a society where generative qualities are present and functioning (p. 643).

Coined by Erik Erickson, the term 'generativity' refers to the investment of time and commitment into the care others; a shift from concern only for oneself toward an interest in contributing to the well-being of future generations (McAdams, Diamond, de St. Aubin, 1997; Peterson and Stewart, 1996; Peterson and Stewart, 1996; Erikson, 1950).

An inherent goal of most societies is the continuing transmission of culture, ideas, institutions, values, and beliefs from one generation to the next. While many theorists have discussed this movement in terms of the stages of life, Erickson was one of the few who examined the stage between age 40 through the end of a person's life, and concluded that older people could play a significant role in forging cultural continuity. McAdams and de St. Aubin (1992) further argued that "cultural demands, generative inner desires, generative commitments, and a belief in self and species, were important" (as cited in Peterson and Stewart, 1996, p. 21), and that this experience is a significant part of the experience of older people.

Extending beyond the scope of generativity is "grand-generativity," when the aged assume a direct responsibility for maintaining the world for future generations. This correlates with such roles as grandparent, old friend, advisor, and mentor. Broadly, "it provides a mechanism for perpetuating wisdom, knowledge, and cultural values beyond the here-and-now, a concept that has also been defined as "cultural generativity" (Kore, 1984) (Warburton, McLaughlin, and Penskin, 2006 p. 116).

Therefore, a decline over a time in generativity activities suggests that future generations may not feel a sense of duty to pass on ideas, experiences, knowledge, and skills that would continue movement forward in endeavors deemed important to the continued existence of future generations.

In the context of the space movement, the absence of generativity could certainly hinder future efforts to convey the importance, history, and future prospects for space travel and space settlement. It would be much more preferable for every individual to see themselves as important players within the scope of the global social, political, and economic systems on Earth and in space.

THE BUILDING OF INTERGENERATIONAL PATHWAYS

Concepts and knowledge, creativity and innovation, mobilization, and stabilization begin to carve the paths that we might use to reestablish ties between the older and younger generations.

CONCEPTS AND KNOWLEDGE

We know that education is the key to successfully guiding the next generation into the future of outer space; which of course means that formal education is a top priority within all societies. By tapping into the resources of older community members to teach, support, and mentor younger children and young adults in a formal educational setting, both young and old would benefit by sharing and learning from one another.

In addition, learning and education are a life-long commitment for both the youngest and oldest members within all societies. Education extends beyond the boundaries of the classroom, and thus its "delivery system" can be said to be both informal and formal in scope. This is where "learning that transforms" comes into play; for it is often through experience that profound insights come about. We must not forget that many great thinkers were raised in impoverished conditions with no formal education, but they still became great contributors to moving ideas forward.

CREATIVITY AND INNOVATION

Creativity and innovation are often fostered through learning that leads to thinking beyond the scope of the box. Building upon teaching, learning, and experiences that foster learning which transforms, both younger and older generations must continue to learn from one another. In Young-ming Tang's article "Fostering Transformation through Differences: The Synergic Inquiry Framework," the notion of synergic inquiry is defined as, "a transformative framework that provides conditions and context, as well as catalysts for cultivating our human capacities for problem solving and creating change and transformation" (Tang, 1997 p.18). It is a framework that allows for free-flowing thinking and learning to be conducted in a learning environment that may seem unconventional, but has implications that reach beyond traditional book learning.

We already engage in a more contained and structured way of learning when, for example, international students visit the United States. While continuing their academics in a traditional school setting, taking them from the comfort of their homelands and implanting them in a way that challenges their own worldviews can lead to such a framework. While NASA has an on-going international learning program, by targeting only the best, the brightest, and the young, it is limiting its scope of what can be learned and shared amongst all young and older learners; which suggests that the scope should be broadened.

MOBILIZATION

Maintaining society's on-going commitments is the role of its institutions, and one of the most important functions is institutionalized support for relationship building between generations. This may include, but is not limited to, policies, funding, and social support; whose goal is to bring people, groups, and nations together. I refer to this as building pathways.

No longer is it appropriate to generalize and alienate one group from each other. Thus, inhibiting myths need to be broken down and new foundations established to bring young and old together to contribute to growth both on Earth and in outer space. All members of society will be cherished as important to the societies in which they live, work, and grow, and will be viewed as contributing members within the context of a community that extends beyond Earth and into the solar system.

STABILIZATION

Re-visioning the old ways and establishing new and flexible ways of learning and understanding will be pivotal to our future in space. Teaching, mentoring, and leadership are part of our present day commitment to bringing young and old into greater harmony.

Effective stabilization requires trust, which is a necessity if we are going to work together. Trust fosters generativity activities, trust creates understanding, trust allows for growth to occur, and trust enables communication to allow for the free-flowing transmission of ideas and experiences among and between generations. In space, trust will be necessary because there may be billions of miles between those on Earth and those on colonies off the planet.

WORKING TOWARDS "IDEALS"

Successfully living, working, and raising families in outer space will require us to understand the needs of all members within a society, and to develop ways to tap into every generation as a critical resource that can contribute meaningfully as we begin to venture into the universe. The young, 0-18 years of age, and the old, 65-85+ years of age, have in recent generations become separated or estranged from one another; but I envision a need for these groups to come together through intergenerational pathways that will bridge these gaps to help shape, create, motivate, and sustain a thriving space program for generations to come.

Transformative learning involves active engagement, sees inclusiveness pertaining to all members as necessary, and does not discriminate. It will play an important role in bringing two generations of people together, uniting people of varying abilities, social classes, genders, and educational attainment. Transformative learning challenges each individual to examine long-held world views, encourages societies to be more inclusive and responsive to all citizens, and does so by creating new ways of teaching and learning that are seen as a life-long endeavors.

So why is it necessary as we head into outer space?

First, our society has allowed gaps and separation between the old and young to form; which has led to the under-utilization of both groups. But it's going to be impossible to continue to ignore these groups in the wake of their continued exponential growth in the near future. This is especially pronounced as older adults look for ways to engage in generative activities that provide meaning, skills and knowledge that contribute to the continued growth of a new generation.

Second, society needs to find a way to nurture youth, thereby providing experiences and knowledge that will get them excited about a future in space. Current NASA educational programming promotes learning to foster interest, but there needs to be an accelerated effort to reach beyond normative patterns of communication that focus on television and radio, to include other media and methods. Older adults might provide the basis for new learning modalities that could positively harness the energy, creativity, and innovation of naturally inquisitive youth who view their world as boundless.

Third, a committed working relationship between people on Earth and those in outer-space may be essential to the success of both. Those who prefer to stay Earth-bound, such as myself, and those who wish to explore the vastness of space, need to support one another. Support may come through families, educational institutions, politics, economics, medicine, and spiritual guidance.

Finally, the magnitude of the work to be done is nearly limitless, and the potential of people all ages to contribute to this great effort cannot be underestimated. We can imagine that in the decades, if not the centuries to come, that each and every person on Earth and in space has meaningful work to contribute to the enduring success of human communities on Earth and beyond. All individuals on a global and trans-global scale may become valuable contributors to the space program; and while those with relevant professional knowledge are of course essential; fostering new ideas, providing support, as well as teaching and learning will come from many different frames of reference. These qualities will emerge through learning and teaching in the home, in school, outside the classroom, through mentorship programs, and through many other forms.

If we are to succeed in the development of communities in space, a unified space community will surely listen and work together well when all voices are heard and all ideas are brought forth, and when training, knowledge, and the opportunity to participate and contribute are available for all, no matter how old, or how young.

•••

CATHY L. WONG, PH.D.

Cathy L. Wong, Ph.D.. has been teaching sociology since receiving her M.A. in Sociology at the California State University Sacramento in 1997 and is currently at the California State University Stanislaus in the Department of Sociology/Gerontology where she has been a full-time lecturer for the past six years. She obtained her Ph.D. in February, 2009 from the California Institute of Integral Studies with a concentration in Transformative Learning and Social Change. The focus of her academic work has been on transformative learning, gerontology, race and ethnic relations, and international and transracial adoption.

REFERENCES

1. Atchley, R.C., and Barusch, A.S. (2004). *Social Forces and Aging: An Introduction to Social Gerontology*. Tenth Ed. Thomson Wadsworth.
2. Baumgartner, L.M. (2001, Spring). "An Update on Transformational Learning." *New Directions for Adult and Continuing Education*, 89, 15-24.
3. Clark, M.C. (1993). "Transformational Learning." In S.G. Merriam (ed.), An Update on Learning Theory. New Directions for Adult and Continuing Education, 57. San Francisco: Jossey-Bass.
4. Cross, B. (2006. "Sowing Inspiration for Generations of 'Space Adventurers.' In Krone, B., Mitchell, Ed., Morris, L., and Cox K., *Beyond Earth The Future of Humans in Space* (pp. 105-113). CG Publishing, Inc., Burlington, Ontario, Canada.
5. Erikson, E.H. (1950). *Childhood and society*. New York: Norton.
6. Freire, P. (2000). *Pedagogy of the Oppressed*. (Revised 20th Anniversary Edition). New York: Continuum, 2000.
7. Hagestad, G.O. and Uhlenberg, P. (2006). "Should We Be Concerned About Age Segregation?" Research on Aging, Retrieved March 19, 2007 from http://roa.sagepub.com/cgi/content/abstract/28/6/638 at California State University Stanislaus.
8. Hooyman, N.R. and Kiyak, H.A. (2005). *Social Gerontology A Multidisciplinary Perspective*. 7th Edition. Allyn and Bacon: Boston, New York, San Francisco.
9. Keyes, C.L.M. (1998). Social Well-Being. Social Psychology Quarterly, 61:2, 121-140.
10. McAdams, D. P., Diamond, A., and de St. Aubin, E. (1997). "Stories of Commitment: The Psychosocial Construction of Generative Lives." *Journal of Personality and Social Psychology*, 72:3, 678-694.
11. Mezirow, J. (1991). *Transformative Dimensions of Adult Learning*. San Francisco: Jossey-Bass.
12. Mezirow, J. (2000). "Learning to Think Like an Adult: Transformation Theory: Core Concepts." In J. Mezirow and Associates (eds.). *Learning as Transformation: Critical Perspective on a Theory in Progress*. San Francisco: Jossey-Bass.
13. Moraitou, D., Kolovou, C., Papasozomenou, C., and Paschoula, C. (2006). "Hope and Adapation to Old Age: Their Relationship with Individual-emographic Factors." *Social Indicators Research*, 76, 71-93.
14. Peterson, B. E. and Stewart, A. J. (1996). Antecedents and Contexts of Generativity Motivation at Midlife. Psychology and Aging, 11:1, 21-33. Tang, Y. (1997, Summer). Fostering Transformation through Differences. ReVision, 20:1, 15-19.
15. Warburton, J., McLaughlin, D., and Pinsker, D. (2006). "Gererative Acts: Family and Community Involvement of Older Australians." *International Journal of Aging and Human Development*, 63:2, 115-137.

Chapter 9

The Next Generation of Industrial Space Facilities: Issues of Management and Operations

Mike H. Ryan, Ph.D.
Bellarmine University
Rubel School of Business
and
Ida Kutschera, Ph.D.
Bellarmine University
Rubel School of Business

Introduction

With completion approaching for the International Space Station, it's now reasonable to consider what comes next. It's always possible to debate the potential of orbiting industrial facilities, and any useful research that might be accomplished within them. If we nevertheless assume that they will be built sooner or later, we can then turn our attention to the management issues that they will require us to consider. These will include an entirely new set of questions that business has never had to deal with before.

    We already know a great deal about what it takes to create successful environments for business operations on Earth, and we know about what it takes to create environments conducive to creative and innovative research. If these necessary characteristics were to be translated to an orbiting facility, what changes would be required in industrial practice, what new operational situations might be encountered, and what new managerial issues and problems would need to be dealt with? These are the questions that we examine in this chapter.

REMOTE LOCATIONS

Working off the planet with a large group of people will yield situations that require competent managers and open minds. Experience obtained about working in remote locations is undoubtedly applicable, but the issues that arise may take new and surprising forms on an orbiting industrial platform.

One of the strongest rationales for creating a facility in orbit is of course that it provides access to either zero gravity or reduced gravity environments. But the advantage that zero g provides for industry is not necessarily conducive to long duration activities by those who live there. While operating in zero gravity does present advantages, such as reduced stress on the body, the ability to overcome certain physical limitations, and even a heightened sense of psychological well-being, research to date (See: J. C. Buckey, Jr., Space Physiology) suggests that the long-term negative consequences of working in near Earth orbit can also be more problematic than helpful. Hazards include issues such as calcium and bone loss, exposure to radiation, loss of muscle strength, potential cardiovascular changes, and prolonged motion sickness. Therefore, determining what needs to be done in orbit compared with what might be done on Earth will be critical to the viability of any orbiting commercial facility. If the activity could just as well be conducted on Earth, and presumably at lower cost, in greater comfort, or without developing new management or research techniques, then the use of an orbiting facility simply wouldn't make sense.

We anticipate, however, that space-based research and manufacturing will provide distinct and specific advantages that eventually will make it commercially viable. Therefore, a serious evaluation of the management issues that we expect to develop in orbit will sooner or later become essential to the decision making process for potential space investors. While space is actually not as remote as many places on Earth, or as remote as a lunar base, an orbiting industrial facility would nevertheless entail some unique operational attributes, and therefore present some interesting managerial problems.

Places Where People Work Frequently Change — People Do Not

It's common today for military contractors on Earth to operate in a military, hierarchical organizational structure, and initially we expect that those who undertake commercial work in space will be required to accept the same conditions as those currently available to NASA and military workers. But as scientific and military activities in space begin to converge with commercial activities, a significant change will take place in both the organizational structures and the operational considerations.

It's likely that citizens in space will view these assignments much in the same way that expatriates assigned to foreign countries do. Their expectations do not include rigid command structures, unduly harsh physical conditions, minute by minute oversight and control, the absence of free time, or the lack of amenities. So as more civilians move into space to explore business opportunities, facility management and the amenities that will be expected there will change from Spartan and militaristic to approaches more similar to mainstream international business practices.

The presence of these practices is more acute for those who are engaged in creative endeavors, who might otherwise feel that restrictions impair their ability to actually do the job for which they were contracted. People working for private organizations will put up with any number of inconveniences given sufficient incentives to do so; however, there are limits. To get the most productivity from those expensive employees and contractors, we expect significant lifestyle improvements will be provided beyond the traditional mission-driven government outpost.

Further, as more facilities are built to accommodate a broader range of commercial activities, an equally broad base of employee types may be anticipated. Many people would like to go into space for the unique experience it offers, but such sentiments will prove insufficient for putting together a cohesive and productive workforce in the long term. It's far more likely that companies will build orbiting facilities that provide conveniences similar to those available in remote locations on Earth, such as remote oil fields or scientific stations.

Significant research has been already done on the problems and issues related to managing people in remote and difficult locations. Jack Stuster's *Bold Endeavors: Lessons from Polar and Space Exploration* (Annapolis, Maryland: Naval Institute Press, 1996) provides many insights that will certainly be pertinent to large scale operations in space. There will be a point where the expectations of the work force reflect those of other employees anywhere who go to work in unique environments, but it is that period between where working in space is extraordinary and where working in space is the norm that we are concerned with.

We expect a transition similar to that experienced by the offshore oil industry. Drilling operations that are incredibly rugged, difficult, and dangerous have been transformed by safety, environmental and profit concerns into work environments that, while the work itself retains its inherent difficulties, are eased by the amenities of a good hotel and the support team that goes with it. 'Floatels,' as they are called, are large residential and service complexes built for those working on offshore platforms. They house over 400 employees in individual, sound-proofed rooms with bed, shower and work space. Privacy, security, 24/7 food service, medical care, recreational areas and amenities such as Internet access are all provided, and supported by additional employees who are hired and trained for those jobs. Floatel residents are more productive and better able to tolerate the difficulties of working long hours at very hard jobs, and space facilities should be expected to incorporate the same lessons.

Basic operational considerations must not be overlooked just because the industrial facility is located in orbit, on the moon or anyplace else. In space, some of the more interesting and potentially problematic long-term issues include creating an effective operational culture, and establishing effective practices for the management teams that control or operate space facilities. Practices such as time determination, employee selection and employee morale are crucial to long-term success.

PRACTICALITIES OF TIME DETERMINATION

It is human nature to attempt to bring a sense of normalcy even to the most unusual circumstances. We expect that working patterns similar to those that people are accustomed to will most likely develop for space-based facilities as well. The development of shifts and working schedules will also help to ensure optimal productivity, and in addition people will expect down-time beyond just rest or sleep to help them regenerate and get reenergized for their work.

Work may be expected to proceed on a 24/7 schedule; as the capital investment in facilities will be so high that almost any other approach may be unacceptable. Maintenance and other station-keeping chores would also certainly require around the clock activity. But even apparently simple shift planning decisions take on a measure of complexity; as the location of a space-based facility might be geosynchronous (stationary over a single point on Earth), or in an orbit that passes over many different locations during each of its orbits. Depending on the location and purpose of the space-based facility, a decision must be made as to which time zone or zones the space-based operations will adopt. Do certain day/night patterns need to be followed? Or would it be best to follow a schedule that is not linked to any Earth-like timetable? These decisions become even more complicated if the facility supports multiple businesses that happen to be based in different countries or time zones.

Traditionally, standard time decisions are determined according to the location of the facility, with staffing adjustments based on customer needs. So there are precedents for using different time zones for business purposes. But there are also precedents for pay adjustments based on night shifts or similar considerations; however, pay differentials alone do not ensure safety or operational efficiency. The problems of day sleepers/night workers are fairly well known, and not everyone can or is willing to make such lifestyle adjustments. Accommodating the needs of space-based, 24/7 employees in terms of support, noise abatement, and activities for free time will be an increasingly critical component for managers of space operations.

ISSUES RELATED TO EMPLOYEE SELECTION

Employee selection is an important issue for every company. Not everyone who wishes to work at night, for example, is able to do so safely or successfully. And some characteristics that might be ignored or neglected in other situations can become potential safety or operational issues when individuals are posted to remote locations, or to space.

And even with training, support, and ongoing oversight, some people simply don't make the necessary adjustments when working abroad. Space-based facilities will be the same. Some employees will not work well, some will not make the necessary adjustments, and some will discover that their needs and expectations are not well served by working away from Earth.

Employee selection is a problem inherent in all business operations. Given the unique nature of space-based operations, the importance of employee selection therefore gets upgraded from high to absolutely critical. Employee preparation costs will be significant, and in addition to the potential financial losses, choosing the "wrong" person to join an in-space team could lead to disastrous or even deadly outcomes.

Psychological stability is essential. Potential space employees will have to prove their ability to work under extreme environmental conditions, and in very close proximity with others; much like NASA astronauts who have to undergo psychological testing and psychiatric evaluations (although even NASA's tests for psychological suitability are not infallible).

Physical health is another major issue. It's likely that medical care will be limited in early facilities, and that they will improve only over time. For example, the effects on bone density due to lack of gravity are well known, and therefore employees will not be permitted to remain in orbit for long periods; unless medically-related side effects can be mitigated or eliminated altogether. [Editors' note: See Chapter 2]

It has been proposed that centrifugal force might be used to compensate for the absence of normal gravity. Large facilities might rotate in some manner to provide all or part of the force necessary to make long duration habitation safe. For many people the rotating space station from "2001: A Space Odyssey" is the image that may come to mind, and is a good point of reference for what many engineers have proposed. Hence, even where the absence of gravity is necessary for industrial operations, an adjacent rotating space equivalent of a Floatel could be constructed for habitation. People would work within a zero gravity environment, but live in a facility providing both appropriate gravity as well as appropriate amenities. But until artificial gravity structures are developed, space-based operations may simply rely on the excellent health of longer duration employees.

Part of the selection procedure could very well consider the willingness to undergo certain preventative procedures that would never be required for a terrestrial setting. For example, while appendicitis is a minor emergency on Earth when medical facilities are usually within easy access, an occurrence in space could have life-threatening consequences. Similarly, simple medical issues like food allergies or high blood pressure that can easily be regulated by medication might not be easily dealt with in space; at least not in early stage facilities, and would therefore constitute a reason to eliminate specific job candidates who might be susceptible to these types of problems.

This suggests that some government employment regulations will be difficult or even impossible for spaced-based businesses to comply with; and 'discrimination' against some job candidates may become necessary and commonplace. The definition of what it means to 'discriminate' itself will therefore need to be re-examined, and so will issues related to jurisdiction. For example, will U.S. Government regulations apply at all in an orbiting industrial facility?

ISSUES RELATED TO EMPLOYEE MORALE

No matter how exciting a new work environment is initially, after some time every employee experiences the development of a routine in his or her work. This is human nature, and in most cases it contributes to worker effectiveness as people master the basics and learn to focus on the subtle differences that are essential. Employees also need the chance to "get out" and do something different on their time off, so the availability of recreational space will be critical to maintaining morale. Space facilities will therefore have to be designed to provide some forms of entertainment. Further, common recreational space does not equate to personal space, and people require both.

Studies of Antarctica expeditioners have shown that the separation from family and friends is one of the biggest social stress factors (See: Harrison, p. 121), so having a place to keep private items such as family photos takes on added importance. It's important for people to have a place where they can be alone. The critical importance of getting a good night's sleep must also be part of the design from the very beginning. Creating greater common space at the sacrifice of less noise in private or sleeping areas will not be an effective long-term tradeoff.

Will married couples be permitted, or even encouraged, to accept positions on space-based facilities? Will families live there, just as they do in expatriate communities worldwide?

In the early stages of space-based facilities it seems unlikely, because employees will need to share spaces like bedrooms and bathrooms with potentially many others; interfering with the solitude and intimacy that couples expect. But reality has a habit of accelerating operational changes, and management expectations that suggest that a prolonged period will occur before people demand privacy and/or private accommodations may also be unrealistic. Lack of privacy, lack of sleep, and lack of just being left alone are all factors that will destroy morale. And, morale will be critical if industrial space facilities are to be operated successfully.

Food and food aroma are critical components in sustaining morale, so managers need to consider food preferences. This in turn also brings up the issue of food preparation. In some cultures food is considered more than just fuel for sustaining the human body. To the extent that food can serve to increase morale, it can also become divisive if an aroma that one person loves is not appreciated as pleasing to others. In the closed environment of a space station olfactory concentrations will increase, and if people of multiple nationalities share a large industrial station, food selection could become contentious. Community guidelines on food preparation could become one of the more interesting social compacts produced on an orbiting structure.

Off shore platforms hire service personnel to provide cafeteria-style food service, complete with an extensive selection and twenty four hour availability. A 500 person facility with the capacity of 3500 meals per week requires a supporting cast of approximately 18 people, and almost two tons of food per day. So very quickly, the expanding employee roster on a space station will probably require the first group of orbiting chefs or short-order cooks.

ISSUES RELATED TO EMPLOYEE MOTIVATION

Motivation may not be the problem, at least initially, because many employees will self select, and they won't be going unless they want to. Those who pass the screening and selection criteria will be among the first select few allowed to work in that environment. Assuming pay and benefits are satisfactory and amenities are sufficient, these people should be largely self-sustaining for a period of time. But as more people move into a station, it will be important for managers to remember that the same principles that prevail on Earth will prevail in orbit; because we are dealing with the same people. Cognition, motivation, and emotion do not change in space.

People attracted to difficult duty environments are likely to have strong personalities and a good sense of their own technical expertise (see Stuster, page 96), but these individuals are not necessarily easy to manage. In remote locations especially, the elements in a manager's motivation toolbox must be used carefully. For example, traditional approaches for punishing bad behavior will probably not be applicable, as most things that one might remove as punishment will be considered essential to survival and maintaining one's sanity, and therefore are unsuited as use for punishment. If even seemingly small things are denied then the result could be considerable demotivation; which then defeats the purpose of punishment.

LONGER TERM ISSUES RELATED TO SPACE-BASED OPERATIONS

The patterns that have governed the world of work on Earth will govern the world of work in space. Given the launch costs associated with moving people into orbit, we assume that the transportation infrastructure will not be in place for some time to allow frequent interchange of large numbers of employees; which means that keeping people in orbit for prolonged periods will be necessary.

And when the number of prospective employees reaches numbers similar to those of a comparable off shore operation, then the facilities to support those people will have to be in place. 'Simple' things like feeding several hundred people per work shift, or providing for appropriate, noise free, safe sleeping areas become essential priorities if business is to operate normally.

Three of the more "interesting" areas for long-term management concern are facility operation, the question of who's in charge, and organizational culture.

In an orbiting industrial facility, it's possible that technical work could be performed under the direction of researchers on Earth, where a group of specially trained technicians would act as the arms, legs, ears, eyes, and hands of those on Earth. This model has been utilized in the American and Russian space programs since the early years of the space age, and it has the benefit of accomplishing experimental work in a space facility that is greatly reduced in size. Under this model, teams of space-based technicians would regularly return to Earth to avoid negative effects from long duration space activity.

The disadvantage, of course, is the potential for lost opportunity, as technical personnel may be less likely to achieve potential new discoveries than those whose work or intellectual curiosity leads them to pursue questions other than those on a prescribed agenda.

Hence, it seems plausible to expect some form of artificial gravity for the habitation areas, allowing for the possibility of longer duration stays for research and analysis; as artificial gravity would allow a greater range of research talent to utilize the space environment than might be possible if only zero gravity were available. The increased variety and number of users would then impact the facility design, and increase the likelihood of shared industrial use, making even large orbiting industrial research centers plausible.

While it's possible that there will be numerous orbiting labs, each pursuing its own research, it seems more likely that multiple organizations will share facilities. If they intend to conduct research that they expect to keep confidential for technical, business, or financial reasons, then some provision will have to be made for confidentiality and privacy.

Some form of shared investment and shared returns among the participants could reduce the security problem and enhance the quality of the output through higher levels of collaboration and creativity. Projects that are too large, too expensive, or cross several areas of interest could be supported by firms that may be competitors in some areas, but which also share resources in their combined space efforts. Long-term collaboration could produce significant synergies that might otherwise be impossible given the inherent cost of space-based research and development.

This second model for an industrial space facility would be an environment where researchers do their own work, and are housed on the facility or a nearby habitat for prolonged periods. This model would accommodate research cycles in the realm of two to four years, consistent with similar projects based on Earth. We can imagine a rotating ring where individual research modules are owned or operated by different companies; the equivalent of a space-based research park.

It might be necessary to place the management of the facility in the hands of managers with unique responsibilities and powers. Much as the captain of a ship is responsible for everything that happens on board, so too would be the senior director or manager of an industrial space station. Those in charge would have to know virtually everything that's going on in order to ensure that projects, activities, or experiments did not create unacceptable risks either separately, or due to some sort of unplanned interaction.

Segregating information among individuals is the most common way of dealing with the potential of unwanted data or knowledge transfer. But in an environment where lack of knowledge has the potential to become catastrophic, segregation may not be practical or prudent. Therefore, professional space managers will be bonded, supervised, and constantly monitored to ensure that they have not engaged in any activities that might create a conflict of interest. Station managers will therefore be responsible to an independent board or agency established explicitly for the purpose of facility oversight and information security.

ISSUES RELATED TO CULTURE

Creating a single culture that is capable of supporting all the activities of a diverse space-based operation is a significant challenge. The organizational culture of an industrial space facility must be designed, because the alternative of allowing a culture to develop from the spontaneous interactions of the initial managers and employees will not assure long-term success - technical organizations are notorious for creating cultures that are myopic and limited in scope.

Cultural conflicts could also stem from the interaction of different cultural and national backgrounds. Geert Hofstede, a noted expert in the field of cultural dynamics, suggests that: "Culture is more often a source of conflict than of synergy. Cultural differences are a nuisance at best, and often a disaster." (See: Geert Hofstede at http://www.geert-hofstede.com.) Putting groups of people together from different cultures with the expectation that they will be able to create a satisfactory working culture without intervention and ongoing support is probably expecting too much.

The two factors of organizational culture that may have the most significance are individualism/collectivism and power distance. "Individualism/collectivism" refers to the degree to which individuals are integrated into groups. On the individualist side we find societies where the ties between individuals are loose: everyone is expected to look after him/herself and his/her immediate family. On the collectivist side, we find societies where people from birth onwards are integrated into strong, cohesive in-groups which often include extended families (with uncles, aunts and grandparents), who continue protecting them in exchange for unquestioning loyalty. Power distance is the extent to which less powerful members of organizations and institutions (like the family) accept and expect that power is distributed unequally. This suggests that a society's level of inequality is endorsed by the followers as much as by the leaders.

American culture is highly individualistic and has low power distance, as Americans are typically not comfortable with unequal power distribution. By comparison, Asian cultures tend to be highly collectivistic with high power distance. These obvious differences provide the potential for not only misunderstandings, but in space for serious consequences when command decisions must be made; as cultural differences can create circumstances where emergency situations move out of control simply due to the manner in which various cultures process and exchange information. For example, there are several reported incidents involving Asian cockpit crews, where a copilot deferred to his captain even though it was apparent to the copilot that an error had been made. Hence, specific training will also be necessary.

Creating a new culture that largely replaces the culture of any individual's national origin will be difficult, but likely important to the long-term success of space operations; and managers will need to be trained to capture critical information that might be otherwise hidden, distorted, or ignored due to the cultural viewpoint or bias of themselves or other groups of employees. With increasing globalization, most businesses are already dealing with similar cultural issues. It's common for managers to undertake cultural training to sensitize them to the importance of cultural differences, emotional intelligence, and soft skills, and this type of training will be critically important to the success of any space management team.

WHERE WILL THIS TAKE US?

Will we see large-scale, privately owned commercial facilities in space within the next 10 to 15 years? Our view is that this is unlikely, considering the technical problems related to putting large facilities into orbit. It seems more plausible that large-scale commercial facilities will be established within 20 to 30 years, but only if the opportunities to provide goods and/or services from an orbiting facility make economic sense.

The capacity of the human species to adapt, endure, and succeed in a wide variety of environments across the planet has long been apparent. But it is the human ability to transform harsh environments to make them more suitable for our needs that sets us apart from other species. Human adaptation to space is far less likely than simply creating structures that will make that unique environment more amenable to human requirements. As Buckminster Fuller noted, the strategy is to change the environment, not the humans.

For comparison, consider how humans have adapted to the ocean environment. Large numbers of humans now live and work below the surface of the sea, and others live and work on large platforms attached to the sea floor. In addition, large cruise ships have been designed for full-time habitation by owner-occupants in mobile communities capable of traversing the oceans.

None of these activities reflect perfect adaptation to the hazards of the ocean environment, but each suggests a strategy whereby humans have found a way to work and live in a difficult environment. This desire to make so-called normal living possible is common to the human condition.

The desire to have as normal a life as is possible, combined with a tendency to embrace the familiar, will extend into space sooner rather than later. People like their creature comforts and common points of reference; and that ultimately will mean that space-based facilities, platforms or other work areas will accommodate the needs of those working in them, and not the other way around. People will both expect and demand that space operations conform to the norms for working in any other remote location. After all, one day in the not too distant future, we'll routinely think of space as 'just another place to work....'

•••

Mike H. Ryan, Ph.D.

Mike H. Ryan, Ph.D. is a professor of management at Bellarmine University in Louisville, Kentucky. He received his Ph.D. and master's degrees from the University of Texas at Dallas. Previously, he founded and operated Prometheus Press, Inc. which published Space Business Notes, one of the Internet's first space business journals. Mike Ryan frequently consults with organizations on issues ranging from technology and innovation to strategy and public policy. He has written extensively on a variety of topics related to doing business in space. He is also a fellow of the British Interplanetary Society, a senior member of the American Astronautical Society and a Fulbright Senior Specialist candidate.

Ida Kutschera, Ph.D.

Ida Kutschera Ph.D. is an assistant professor of management at Bellarmine University in Louisville, Kentucky. She received her Ph.D. in organizational science from the University of Oregon and has an MBA from Washington State University. Her current research interests include social cognition and decision making, in particular the impact of cognitive styles and the use of intuition in managerial decision making, as well as management and leadership in extreme contexts. Ida Kutschera is a member of the Academy of Management and the Society for Human Resource Management.

SELECTED REFERENCES

1. Berinstein, P. Making Space Happen: *Private Space Ventures and the Visionaries Behind Them*. Medford, New Jersey: Plexus Publishing, Inc., 2002.
2. Buckey, Jr., J. C. *Space Physiology*. New York, New York: Oxford University Press, 2006.
3. Goehlich, R. A. *Spaceships*. Ontario, Canada: Apogee Books, 2006.
4. Greenberg, J. S. *Economic Principles Applied to Space Industry Decisions*. Reston, Virginia: American Institute of Aeronautics and Astronautics, Inc., 2003.
5. Harrison, A. A. *Spacefaring: The Human Dimension*. Los Angeles, CA: University of California Press, 2001.
6. Hudgins, E. L. (ed.). Space: The Free-Market Frontier. Washington, D.C.: Cato Institute, 2002.
7. Kitmacher, G. *Reference Guide to the International Space Station*. Ontario, Canada: Apogee Books, 2006.
8. Ryan, M. H. and Kutschera, I. "Lunar-Based Enterprise Infrastructure – Hidden Keys for Long-term Business Success," *Space Policy*, Vol.23 (2007), pp. 44-52.
9. Ryan, M. H. "The Role of National Culture in the Space-Based Technology Transfer Process," Comparative Technology Transfer and Society, Vol. 2, No. 1 (April 2004), pp. 31-66.
10. Ryan, M. H. and Luthy, M. L. "Management Architecture: Problems Facing Lunar-Based Entrepreneurial Ventures," *Journal of Space Mission Architecture* Vol. 1, Iss. 3 (2003), pp. 20-38.
11. Stuster, J. Bold Endeavors: *Lessons from Polar and Space Exploration*. Annapolis, Maryland: Naval Institute Press, 1996.

CHAPTER 10

DEVELOPING A SPACE COLONY FROM A COMMERCIAL COMET MINING COMPANY TOWN

THOMAS C. TAYLOR, M.S.
GLOBAL OUTPOST, INC.
AND
HAYM BENAROYA, PH.D.
PROFESSOR, DEPARTMENT OF MECHANICAL & AEROSPACE ENGINEERING
DIRECTOR, CENTER FOR STRUCTURES IN EXTREME ENVIRONMENTS
RUTGERS UNIVERSITY

INTRODUCTION

There are two primary reasons for humans to venture to a remote comet. The first is a commercial desire to develop a valuable resource, such as water. The potential profits could be significant enough to support the creation of a company town, and perhaps eventually a space colony with as many as 10,000 inhabitants. A second, even more compelling reason is that the knowledge gained through exploration, landing, definition, extraction and ultimate control of these objects might enable us to learn how to control large objects that threaten the destruction of Earth.

About 2,225 near-Earth objects are currently known in the 10m to 30m size range. Detailed information exists for a mere 300 of these. It is estimated that there are roughly 25,000 objects larger than 150m in size. It is estimated that only 250 of these are potentially hazardous to Earth. The number of objects larger than 1 km, the size that is capable of causing global scale catastrophe, is now estimated at between 900 and 1230. Around 55% of these have been specifically identified. None of these are known to be on Earth-intersecting trajectories. In the event that one of these objects is identified as being a threat to Earth, mastering techniques to relocate these objects is a key side benefit of the following proposed endeavor.

A mining operation that evolves into a space colony might be one way to accomplish the dual goals of Space exploration and human settlement in Space. This model has been widely and successfully used on Earth with examples such as the Hudson Bay Company, the Alaskan Oil Fields, and mining towns throughout the American West.

"Space Commerce" in its many forms will lead to increased markets, investment capital from sources other than taxes, lower costs, and multiple innovation paths that the government does not have to fund with tax dollars (until they purchase services and enjoy the competitive market savings) (Taylor, et al., 2007, 2008). It is both desirable and inevitable.

And to those who say that the commercial development of space should come after NASA and other space faring nations have "completed" their exploration, let us respond that commerce is the way the world works, and establishing early trade routes will likely be central to the development of Space.

WHAT IS THE BUSINESS CASE AND HOW MIGHT SPACE COMMERCE UNFOLD?

Space development currently moves at a pace that is defined by government expenditures. In contrast commercial development moves at a pace governed by anticipated future profits and market forces.

Twelve years ago, most people within the aerospace community thought that reusable launch vehicles would reduce the cost of launch to orbit by several orders of magnitude. It now appears that only a one order of magnitude reduction is achievable on the current commercial horizon (Citron, 1995). The consequence of this has significance for our envisioned space-water business, because the market price for water creates a commercial opportunity.

Water is currently worth approximately $10k per pound in LEO. However with the introduction of commercial reusable launch vehicles the cost should drop to $1k per pound in LEO. A transport vehicle containing

72,000 cubic feet of comet water (the approximate volume of a Space Shuttle external fuel tank), delivered to LEO is therefore projected to be worth between $4.5B (at $1k per pound) and $45B (at $10k per pound). Delivered to L-1 and GEO, it will be even more valuable.

Hence, in this business concept, water will be sold for human consumption on other off-Earth human habitats such as space colonies and orbital LEO and GEO science, technology, manufacturing, and tourist destinations. It will also be sold as an ingredient in rocket propellants for spacecraft.

In current space logistics, most cargo transport is one-directional, outward only. By contrast manned spacecraft require two-directional transportation. According to current government designs Lunar cargo in the future also appears one-directional. However from a commercial perspective it will be two-directional. Trade routes are usually two-directional. Commerce drives the transport costs lower through volume growth and competition.

In addition to water transport, a new entity, the Comet Logistics Transportation Company, anticipates finding ready customers for cargo transport in both directions. These will include ventures to and from our proposed Wilson-Harrington mining operation and supply circuits and visiting other space habitats on comets, planets, moon, and orbital stations. This proposed new "asteroid trade route" will utilize existing vehicles. Propellant production will serve as its economic basis, and a commercially financed small fleet of propellant supply vehicles will traverse a two-way highway between Earth and off-Earth resource locations.

The commercial endeavor proposed herein is designed to sustain growth for more than 50 years. The business is to be founded on profits from the sale of water, and will be sustained by the opportunity to provide cargo transportation in both directions to support resource recovery from other comets and asteroids. It would be financed by private sector partners and/or by interested nations. The financing method would use the same risk money versus profit techniques used to finance almost every resource recovery development venture on Earth including the 19 new oil fields around the original Prudhoe Bay, Alaska Oil Field. The risk money is minimized to the greatest extent possible, because it is the hardest to obtain. The later profits are used to expand the project after the market for the resource is proven and supporting markets are in place. Government can help accelerate the commerce by various means. After that additional private investment is easier to obtain. This is the exact reverse of the big up-front government budget only method used for space exploration today. Oil is a known resource on Earth, water is a known resource in Space and the risk is more in the doing and less in the finding.

MINING A COMET

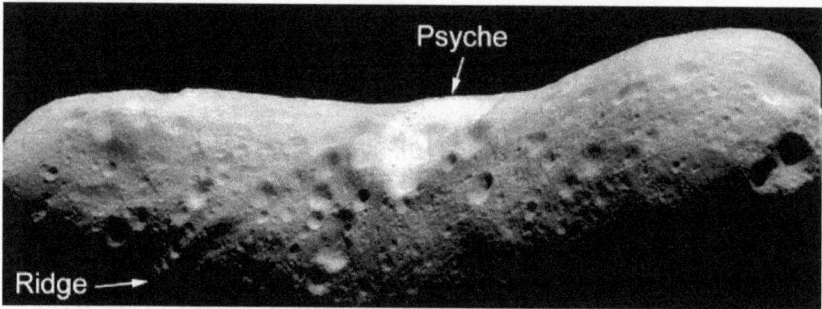

Figure 1. Eros is one of many similar objects within our solar system. It is shaped like a potato and has many impact craters.

Figure 1 shows an example of a burned out comet in our Solar System, one that could contain water ice as well as a number of other resources. This photo shows an object in our solar system called Eros, and as is evident, most objects in space get severely beaten up over millions of years. Many of them eventually look like this, similar to a big potato.

Ridges and groove systems on the surface of Eros are at least several km long, and display topographic relief on the order of 100m (Cheng, 2002). The marked "Ridge" is 18 km long, and extends to the opposite side, suggesting a fracture that extends through the body of Eros. The 5 km crater "Psyche" is also noted.

While Eros is not actually a comet, our hypothetical focus is on a comet (not pictured) called Wilson-Harrington (1979) or Comet 107P/Wilson-Harrington. It has had several other names, partly because astronomers found it, but then lost track of it for a while. Little is known about this comet that would distinguish it from other comets, except for the fact that it is also designated an asteroid dubbed 4015 Wilson-Harrington. For more information see http://www.answers.com/topic/107p-wilson-harrington

Some comets such as Wilson-Harrington contain water which evaporates or emisses to form the comet's distinctive cloud, or tail, in the vacuum of space. Wilson-Harrington ventures through our part of the Solar System every 4.29 years. Sometimes it looks like a comet with a vapor tail, but at other times no tail is evident. Comet astronomers (Hale, 2007) suggest that it is probably a comet with an extinct or, at best, a dormant nucleus. Reports indicate an approximate diameter of 2.4 km, a rotation period of 6.1 hours, a geometric albedo of 0.05, and an orbital period of 4.29 years. Perihelion distance is 0.992 AU, and orbital inclination is 2.8 degrees (Osip et al., Icarus, 114, 423, 1995 from the Minor Planet Center's Catalogue of Cometary Orbits).

The authors suggest that the absence of a tail may sometimes be the result of a sealing-over process, whereby over millions of years, melting surface materials meld to become a coating that then prevents the further evaporation or vaporization of the core. Comets with tails may simply not be sufficiently sealed for effective recovery of the interior resources of the comet, but Wilson-Harrington might be an example of the sealed-over variety. If that is the case it is ripe for recovery because it has reached the stage where humans could effectively tame the comet and recover valuable interior materials using the mining techniques described below.

THE COLONIZATION AND DEVELOPMENT OF COMET WILSON-HARRINGTON

The Wilson-Harrington comet is large enough to accommodate a town, and it has a number of critical, self-sustaining characteristics. These include a commercially viable enterprise of providing water to other off-Earth settlements. The first workers who mine ice from the interior of the comet will live in the vehicles that transported them and their provisions to the comet. Eventually the inhabitants will develop a crude but evolving living environment inside the ice mass. They will provide one-g by augmenting the natural rotation of the comet with additional artificial rotation.

The actual conditions on the surface and in the interior of Wilson-Harrington are not known. Therefore the first step in the development process is to send a package of instruments to learn more. If conditions are found to be appropriate, the next steps, as shown in Figures 2 and 3, will be

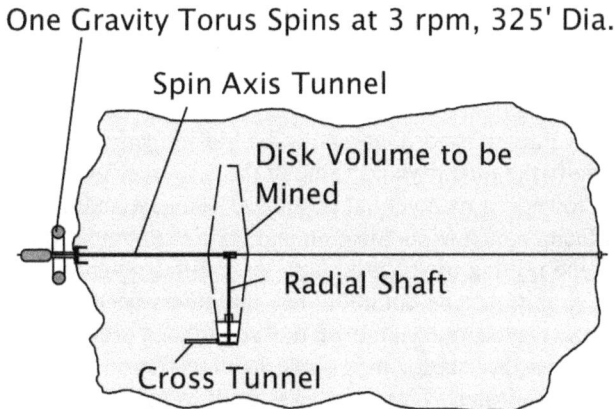

Figure 2. The modest start of mining operations on Wilson-Harrington, with tunnels and shafts bored.

laser drilling or melting a central tunnel along a single axis of rotation, which would lead to the development of the inside of the comet.

After the main centerline tunnel is complete, one end is designated for discarded materials and the other end is developed for delivery of salable water. A system of machines for ice mining and water recovery will collect and transport materials for use by the colonists and for water ice shipment to customers.

If the miners create a 2 km diameter disk inside the ice interior for their mining operations, then about 3% of normal Earth gravity will be created from the natural rotation of the comet at 6.1 hours per revolution. This might support farming. In addition plants could also be cultivated under 100% grow lights.

Most space colonies are not visualized at the beginning stages of development, but only later, after the hard initial work is completed and more expansive facilities are finally in place. The 1981 movie, *Outland*, a science fiction film written and directed by Peter Hyams and starring Sean Connery, does a very credible job depicting the difficult social environment and challenging economic forces that may be at work in an early stage space mining community. Hyams depicts a challenging living situation, hard work, and a stressful social environment.

THE COMPANY TOWN

At the Prudhoe Bay Oil Field development in Alaska, 12,000 construction workers labored for five years to construct $20B in recovery and pipeline facilities to extract trillions of dollars worth of oil. However before any oil actually flowed (Taylor 1981) many families and tourists went to the Alaskan North Slope. As more and more facilities were put in place to accommodate those people, a company town emerged.

Similarly, in the American West, towns were established where something of value existed that could be developed as an economic resource. Initially the quantity and value of the resources available dictated the size of the town, but its potential to survive as an expanding community depended on factors such as its location and the development of secondary industries. Some mining towns grew into large cities when they expanded beyond being simply mining communities, but others became ghost towns when the mines played out. One of the co-authors grew up in eastern Colorado in a town that started as a cattle loading station on a spur of the Transcontinental Railroad. The siding was built to handle cattle that were grazed nearby and were thereafter shipped to eastern markets. The cattle business was good on the J. L. Brush ranch and the town grew to include irrigated farm land. Later the town underwent an oil boom and a Ballistic

Missile Silo construction boom. Today the cattle operations are feedlot operations and constitute a mere 20% of the economy of Brush, Colorado.

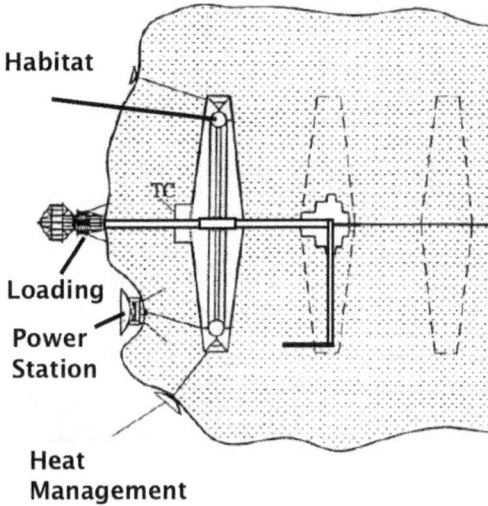

Figure 3. This figure depicts the start of the process of mining and selling of water, thus establishing the economic basis for a company town on Wilson-Harrington.   A one-g rotating wheel permits early normal living on far left and later in the spinning habitat inside the comet, after the water resources are recovered and sold.

HOMESTEADING

What could possibly encourage someone to want to live on a comet far from our home planet Earth? Figure 4 depicts "land" as a homestead, and the economic model suggests that the land increases in value over time as the economy of the comet town develops. Just as in the history of human civilization the homesteading comet families develop their assets in a new economic frontier.

Figure 5 depicts the section details of a One Gravity Wheel (Grandl and Germano, 1995) inside the excavated ice cavern. It shows the end section of a company town ring, where families live in one-g. Some farm their "land". They consume part of the food and sell the rest. Others work in the ice mine.

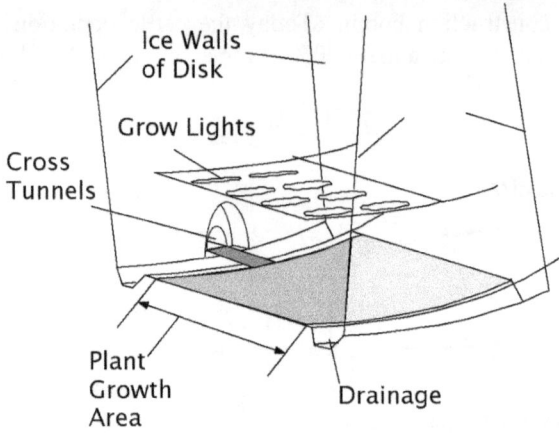

Figure 4. "Land."

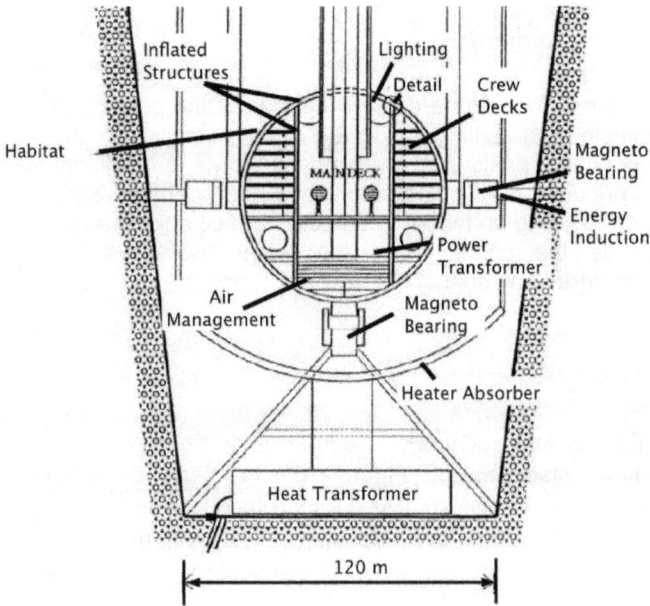

Figure 5. The One Gravity Wheel at 2 to 3 rpm.

Figure 6. Cutaway view of the One Gravity Wheel

ONE WAY TO THE STARS

When the ice and water resources are nearly exhausted, the resource recovery equipment can be moved to another comet. The fully developed town meanwhile could be outfitted with advanced propulsion systems strong enough to accelerate the entire comet shell and its contents to the nearest "Class M" planet, thus enabling exploration beyond our own solar system (Roddenberry, 2000). This means that kindergarten children alive today could eventually be the first humans to leave our solar system in search of other life. Their great-great-great ... grandchildren may one day orbit another star.

CONCLUSION

This chapter offered a commercial mining perspective on the evolution of a space colony. We examined company town operations, camp consumables, and commercial sales of propellant from a comet mining operation. We also examined commercial efforts to develop a water source using an off-Earth body. We proposed a subsequent effort to create an off-planet population that will ultimately be capable of becoming a self-sustaining, financially viable space colony of 10,000 people who thrive without support from Earth (Taylor et. al, 1995).

The co-authors come from different backgrounds. Taylor's is in remote base resource recovery in extreme Earth environments (Taylor, 1975). Benaroya's is in extreme environment modeling and habitation design (Benaroya, 2001, 2002).   In this chapter we combined our perspectives to outline a plan to mine the ice from within a comet over a 40-year period.

We propose that at the end of this period, when the ice is nearly exhausted as a resource, a portion is saved for propulsion of the comet remnants.   With 10,000 people living in a safe, one-gravity, radiation-protected environment "in-board" (literally inside the comet), the comet is to be propelled to the closest star system containing a possible habitable planet.

We realize that even after decades invested in developing advanced propulsion technology, it may still require five generations for space colony inhabitants to complete the trip.   Nevertheless, we anticipate that many aspects of this home-away-from-Earth will appeal to and attract homesteaders who are looking for an "out of this world" existence.

•••

THOMAS C. TAYLOR, M.S.

Tom Taylor is a Commercial Space Entrepreneur and an inventor. Tom is a graduate of Colorado State and Stanford Universities.   He holds 15 U.S. Patents. In the most recent 30 years, Tom has helped found and has worked in the trenches as a startup team member in over 20 commercial space companies. Four of those became successful enough to raise a total of over $1.2B in private equity. The latest startup group, Lunar Transportation Systems, Inc., focuses on the development and logistics of transportation to and from the Moon's surface in support of mining and commercial development.

In a former career, Tom worked 15 years as a Professional Civil Engineer and heavy construction supervising engineer. He has five years of experience working in remote environments such as in the jungles of Thailand. Tom served as a second lieutenant in the U.S. Army Corps of Engineers. He also worked on the oil field construction on the North Slope of Alaska.

HAYM BENAROYA, PH.D.

Haym Benaroya, Ph.D. is Professor of Mechanical and Aerospace Engineering at Rutgers New Brunswick. His degrees are from The Cooper Union in New York and the University of Pennsylvania in Philadelphia. After graduating from Penn, he joined Weidlinger Associates Consulting Engineers in New York in 1981. He went to Rutgers in 1989. His professional interests focus on the study of structural dynamics and probabilistic modeling, with applications to offshore, aircraft and lunar structures. He is interested in space exploration and settlement, education, and defense policy issues. Currently he is the Director of the Rutgers Center for Structures in Extreme Environments. Professor Benaroya is the author of two research monographs on offshore structural dynamics and two textbooks on vibration and probabilistic modeling.

ACKNOWLEDGMENTS

The authors would like to thank A. Germano for his contribution to the original paper, IAA-95-IAAA.1.3.03, "Commercial Asteroid Resource Development and Utilization," presented at the 46th International Astronautical Congress, in October 2-6, 1995 in Oslo, Norway. The authors would also like to thank Werner Grandl, Architect, Martina Pinni, Space Habitation Architect, Assistant Professor, University IUAV of Venice, Dr. Anthony C. Zuppero of the Idaho National Lab for his initial encouragement, and for his contribution to the original paper and technical data on the comet Wilson-Harrington and his contribution to the Oslo paper. Thanks to Alan Hale, astronomer for the recent technical data on the comet Wilson-Harrington. Thanks to Global Outpost, Inc., the development company that signed a NASA Enabling Agreement in the 1980s and placed a cash deposit for five External Tanks in orbit to continue this space entrepreneurship effort for the movement of humanity off this planet, which is called "space commerce" by some.

REFERENCES

1. Benaroya, H., 2002, H., L. Bernold, K-M Chua, "Engineering, Design and Construction of Lunar Bases," *J Aerospace Engineering*, Vol. 15, No. 2, April 2002, 33-45
2. Benaroya, H., "Prospects of Commercial Activities at a Lunar Base," *Solar System Development Journal*, (2001) 1(2) 1-19 (ISSN 1533-7405)

3. Cheng, A., Applied Physics Lab, NASA Workshop on Scientific Requirements for Mitigation of Hazardous Comets and Asteroids, July 2002

4. Citron, B., Kistler Aerospace Corporation Business Plan, June 1995.

5. Fletcher, S., President Reagan's Commission on Space, National Commission on Space, Washington, DC, 1988.

6. Germano, A., Grandl, W., Taylor, T.C., Zuppero, A.C., "Commercial Asteroid Resource Development and Utilization," in IAA-95-IAA.1.3.03, Oct. 1995, Oslo, Norway, presented at the 46th International Astronautical Congress, in October 2-6, 1995.

7. Hale, A., Astronomer, emails and personal communications, 2007.

8. Hyams, Peter, writer and director, *Outland*, 1981. This science fiction movie starring Sean Connery depicts a huge mining operation on the moons of Jupiter. Families are not present, and off-duty time is like a Western cowboy town with a Marshall who keeps order.

9. Pinni, M., Space Habitation Architect & course work, Assistant Professor, University IUAV of Venice, Italy, ARTEC, Dorsoduro 2196, 30123 Venezia, Italy

10. Roddenberry, G. "Class M planet" is the term Roddenberry used in Star Trek, which denotes a planet with capability to support life in some form.

11. Szabo, N., *Comet Mining -- An Overview*, Copyright 1993, 1994 http://szabo.best.vwh.net/comet

12. Taylor, T.C., Personal Experience 1975-1979, Civil Engineer – Construction Supervisor, North Slope Oil Facilities, Prudhoe Bay, Alaska, various technical papers on remote harsh facilities construction environments (1981).

13. Taylor, T.C., Zuppero, A.C., McFadden, L. "Resource Recovery Facilities for Earth Approaching Comets," in *SPACE 94,* ASCE, Feb. 1994.

14. Taylor, T.C., Inventor, USPTO Torus U.S. Patent No. 6,206,328, "Centrifugal gravity habitation torus constructed of salvaged orbital debris," Issued Mar. 27, 2001.

15. Taylor, T.C., "Economic Public Private Partnerships for Development," Session CT108, STAIF - 2008, Feb. 10-14, 2008, Albuquerque, New Mexico

16. Taylor, T.C., Kistler, W. P., Citron, B. "Innovating Public Private Partnerships and Dual Use Technology,", 58th Int'l. Astronautical Congress, Hyderabad, India, www.iac2007.org 23 Sep 07

17. Taylor, T.C., "Lunar Commercial Mining Logistics, " Session B06, STAIF - 2008, Feb. 10-14, 2008 · Abq, NM

CHAPTER 11

STRATEGIC IMPLICATIONS OF ENERGY FROM SPACE

EDWARD B. KIKER

INTRODUCTION

Space-based power sources can help to protect the environment by providing pollution-free energy, and thereby transform the lives of people on Earth for the better in many ways.

Energy from Space will not necessarily be cheaper than what we have today, but as the technologies and capabilities mature, Space-based energy will become more dependable, and thus will stabilize large sectors of the world economy. Strategically, nations and economies will become more independent both politically and economically. This transformation will probably be very gradual and will likely occur over the next thirty to eighty years.

Even if renewable energy sources could provide all the transportation and electrical energy we need, oil, coal and natural gas will continue to be in competitive demand as manufacturing feedstock. When those cease being primary energy sources for transportation and the generation of electricity, they will no longer be significant causes of international conflict and will last much longer.

Energy from Space will expand people's worldview. People will be looking outward more than before, in part because they will recognize Space as the source of their energy.

SOURCES OF ENERGY

Possible sources of energy from Space include Earth-based ground stations that transmute sunlight to electricity during daylight hours, solar power satellite (SPS) systems in geosynchronous orbit that beam power to Earth by microwave and/or laser, solar power systems on the Moon that beam power to Earth by microwave and/or laser, and fusion energy based on the helium-3/deuterium fuel cycle that will utilize helium-3 mined on the Moon.

These various sources have the potential to provide virtually limitless and uninterruptible energy. Implementation will require: (1) the political will to do so, and (2) significant capital resources.

As additional Space-based power stations are built, Space-based power can then be "exported" to any country as either a commodity for sale, or as foreign aid. The People's Republic of China, Russia, Japan, the Republic of South Korea and other nations are all working on variations of Space-based energy technologies.

There are also, of course, many Earth-based energy sources, some of which are almost completely untapped, so Space-based power systems must be considered within that larger context. Oil, hydrothermal, hydroelectric, coal and other sources of energy will continue to be used where they are economically competitive and/or are state-subsidized.

Energy required for vehicles will be largely hydrogen/oxygen combustion or electric battery. Energy required by industry and residences will be largely derived from:

(1) Electricity from fission and fusion installations,
(2) Ground-based solar systems, or
(3) Space-based solar power receiving antennas.

Some residential and industrial electricity and heating will come from private solar photovoltaic panels or solar heat engine units integrated into buildings, and from wind electric power where practical. Solar photovoltaic power, solar heat engines, and Space-based solar power receiving installations may become common in desert areas transporting power to users via the existing electric grid.

Solar power stations in orbit and on the Moon will require mechanics, cooks, barbers, pipefitters, engineers, and many other people in other occupations. These workers will be living in Space habitats, and their respective preferences will affect the way their private and communal spaces are built and decorated. Their living and working spaces will not look the same as the sterile environments that are so often depicted in movies.

A point perhaps more important is that people from many cultures and nations will be working together in a dangerous environment, and this may have strategic implications itself by showing that diverse people can, when they must, get along and build a common future. [Editor's note: See Ryan and Kutschera]

They will have spare time for reading and the arts, and the inspiration brought about by humans living in Space will affect those living on Earth as well. Today's schoolchildren living in diverse places ranging from New York to Beijing know about polar bears and many other things in places far away from their respective home cities. In the future they will become familiar with Space solar power satellites, Moon bases, and zero-gravity ballet.

ENERGY FROM SPACE

Twenty-first century military strategy is popularly considered to be based on the need for oil for running industry, vehicles, and military activities. Many current wars, indeed, have been referred to as "oil wars," by both opponents and supporters, and the repercussions affect every nation. The burning of oil and coal are likewise a point of environmental contention throughout the world, because the resulting pollution affects human health, building preservation, ecology, and the global climate to name a few.

What every country needs in today's competitive international arena is not necessarily oil, but energy, and there are extensive new sources that are virtually untapped. Among those are:

1. GROUND-BASED SOLAR INSTALLATIONS
Photovoltaic: passive solar panels that directly transmute solar energy to electricity, and which may track the movement of the sun.

Heat engines: Solar mirrors which concentrate solar energy onto small structures, boiling a liquid to steam which in turn drives a turbine, or forces movement of a piston, which in turn produces electricity.

2. SOLAR POWER SATELLITES
These are called "Glasers" after their inventor, Dr. Peter Glaser. Glasers will be in geosynchronous orbit, beaming energy to Earth via receiving antennas (rectennas) by microwave at half the intensity of normal sunlight, and with a beam frequency that lets it penetrate clouds while being harmless to humans, wildlife, and aircraft.

Laser transmission is an alternative to microwave transmission, and each method has its strengths and weaknesses. Microwave energy can penetrate all weather but the beam spread can be considerable. Laser energy often cannot penetrate cloud but can have a much tighter beam with less energy loss. Some Glasers may have both technologies, using one or the other depending upon the atmosphere that needs to be penetrated.

3. LUNAR SOLAR POWER

Solar power from solar farms located on the nearside of the Moon will be built. They will be constructed largely of lunar materials. Power from them will be beamed to Earth either by microwave or laser.

4. HELIUM-3 FUSION

Clean nuclear fusion can be produced from the fusion of deuterium and helium-3. Helium-3 has been identified in lunar rocks in amounts that could provide all of Earth's electrical needs for perhaps four thousand years, while deuterium is plentiful in Earth's oceans.

The basic technologies currently exist for the first three solar power sources which are: Ground-based Solar Installations, Solar Power Satellites, and Lunar Solar Power to be realized, so creating these systems is largely a matter of funding and execution. All three technologies will continue to be further refined. And while creating a network of solar power satellites would require significant launch capability, more than we have at present, it can certainly be done.

Lunar installations could be built in part from lunar materials, thereby reducing the need for launch capability. We are currently developing fusion technology, and the capacity to handle the temperatures involved in deuterium/helium-3 fusion reactions seems to be a matter of engineering development, and could be realized within the next ten to forty years.

As these new energy sources come on-line and are improved, they may become cost-competitive. Furthermore, once they become available, finite resources such as oil and coal will receive a renewed lease on life - in that they will last longer than they otherwise would.

OIL AND NATURAL GAS

In addition to energy sources from Space, we will continue to improve the technologies used to exploit existing resources. We will continue to search for oil and natural gas; improve extraction and refining technologies; and seek ways to reduce carbon dioxide, mercury, and other emitted pollutants.

We will also continue to improve our ability to extract oil from the bitumen found in extensive reserves of tar sands and oil shale. Estimates of total recoverable oil from these sources is around three trillion barrels.

Research is also proceeding in the search for oil and gas located in the Earth's deep interior. Geological exploration and theory development indicates that much of what we have traditionally regarded as "fossil" fuels may not be of fossil origin at all, but have been created from hydrocarbons incorporated in the original formation of the Earth and which have since been migrating towards the surface and are now trapped under impermeable igneous or sedimentary rocks, or under permafrost. These gases and oils are sometimes referred to as "abiotic oil" and debate continues on this subject. The association of helium with many oil and gas resources tends to support this theory. Thomas Gold's *The Deep, Hot Biosphere* is one exposition of this theory. Already several deep wells have been drilled to a depth of over 6 kilometers in igneous bedrock in Sweden and Russia. It appears these efforts have been successful. In the deep ocean, the combination of low temperature and high pressure keep gases contained in ice.

COAL

Worldwide coal deposits are vast, however many contain high sulfur concentrations. Sulfur and mercury-bearing coals cause environmental problems. The technologies of coal gasification and extraction of synthetic oils from coal have improved greatly in recent years, and will continue to do so as emission scrubber technology continues to improve.

NUCLEAR FISSION

Nuclear fission plants continue to be built albeit at a slow pace. Although about 80% of the energy in France is generated from fission, for example, considerable concern surrounds the treatment and disposal of radioactive waste from these reactors. A single nuclear accident, such as that at Chernobyl, could wipe out all the economic advantages of twenty or more successful nuclear plants, and could have devastating impacts on cities, populations, and farmlands. Nuclear fission has a future, but most researchers would prefer to see a shift to a non-radioactive nuclear program such as fusion using the deuterium/helium-3 fuel fusion cycle.

EARTH-SURFACE SOLAR

Solar panel technology is improving at a rapid rate. For example, scientists recently created a carbon fiber nanotube material that absorbs more than 90% of the sunlight that falls on it, and this material will be applied to photovoltaic panels, reflectors, and heat engines. As municipal codes and industry standards advance, in the near future solar panels will be incorporated into many types of buildings.

WIND-ELECTRIC

Local availability of constant or frequent wind within useful velocities is a limiting factor in wind-to-electricity conversion. However, much more energy can be generated with wind-electric power than is currently being performed. The method of energy production via the use of wind is growing at a rapid pace.

HYDROTHERMAL

More electricity will be generated from hydrothermal sources in the future. Geologically active areas such as Yellowstone Park may be good sources for this type of energy. A limiting factor is that this technology must be carefully implemented so it does not cause geologic disruptions.

HYDROELECTRIC

Most of the Earth's rivers that can accommodate hydroelectric dams have already been dammed. There will be a few more built. Existing dams and generating stations will be modernized to improve electricity yields, however by and large hydroelectric energy production has neared its maximum.

OCEAN THERMAL ENERGY CONVERSION (OTEC)

This relatively new technology obtains usable energy from warm ocean surface water. Several test plants have been built and there is considerable interest worldwide. Its use is expected to grow quickly. By nature, this source of energy is limited to coastal regions within about thirty degrees north and south of the equator.

TIDAL

Compared with other energy sources, little effort has been made to tap tidal energy. A major obstacle is keeping the tidal pontoons and/or other equipment safe from storm surges. However, test sites and designs for harnessing tidal energy are proceeding.

METHANE HYDRATES

The economic feasibility of obtaining methane gas from methane hydrates on the ocean floor and in permafrost regions is currently being examined. It might prove difficult to separate methane from the soils and sediments, but it can be done. Any mining activities will have to be done very carefully to ensure the hydrates, on their approach to the lower pressure of the surface, are completely contained because methane hydrates boil off as methane gas. Methane gas is a potent greenhouse gas, therefore great care should be taken to prevent it from venting into the atmosphere.

BACTERIAL BIOMASS CONVERSION

Methane, diesel, gasoline, oils, alcohols, and a variety of fatty acids can be obtained by bacterial conversion of cellulose plant fiber and animal waste. A few examples of the materials used are: switchgrass, creosote bush, farm plant waste, animal droppings and abattoir refuse. India has been producing methane in digesters for many years at the village level. Some bacteria are being biologically re-engineered to produce a greater variety of products. Eventually little cellulose or animal waste products will escape bacterial conversion.

PROSPECTS FOR THE FUTURE

Should we now go directly to full-scale development of the four Space energy sources which are: 1) ground-based Solar Installations, 2) solar Power Satellites, 3) lunar Solar Power, and 4) Helium-3 Fusion? The answer is not clear-cut.

We could immediately build very large ground-based solar systems utilizing both photovoltaic and heat engines. Because these technologies are fairly well understood, with additional research and testing of the necessary hardware one each proof-of-principle lunar solar farm and solar power satellite could be built in the near future if we have the will to do it.

The first Glaser will be small. It is likely the first tests will involve transmitting power to a remote site, perhaps a remote island in the Pacific Ocean. The initial lunar solar farm will also be small, and will probably be set up using robots. Building of an experimental fusion energy plant will need to wait until the technology is further developed. Before proceeding, we need to learn how to handle the high temperatures involved. In the meantime, we should send a helium-3 gas mining robot to the Moon as soon as possible to begin development of the technologies for helium-3 recovery.

As these various technologies mature, they will, over time, take on a larger and larger load of our energy requirement. I estimate the final transition will occur sixty to eighty years from today.

In my view, the one resource that will eventually take over the preponderance of energy production is fusion energy using helium-3 from the Moon. The second highest energy supplier will be Earth-based solar and wind power, the third will be solar power satellites, and the fourth will be lunar solar farms.

These estimates are based on the following points. Fission and fusion energy plants have a very small footprint, and fusion energy plants using helium-3 and deuterium are far safer than fission. Therefore, as the technology comes of age, I believe fusion energy will garner considerable support.

While Earth-based solar and wind power are the easiest to develop, there is considerable resistance when the locations of prospective sites are made known to the public. Many people do not want the apparatus to be obtrusive, so aside from widespread solar installations on rooftops and integrated into windows, the necessarily large solar and wind farms must cover vast areas of deserts and plains. However, there already is a fair amount of opposition from environmental lobbies which oppose locating these systems in deserts.

Wind farms on the plains can provide a lot of energy, but wind is not always reliable. Furthermore, current transmission capabilities are not adequate to enable large-scale power transmission from the plains to far away major coastal population centers.

Solar power satellites may never be developed as anything less than small, single purpose units. One reason is for this is that the very large Glaser structures that would be required to run major metropolitan areas will not only take far too much energy to get up and in place, but they are also likely to generate considerable space debris as meteorites strike them and thereby cause debris to litter Earth's geosynchronous orbit. Because the geosynchronous orbit is so important to communications and other purposes, such debris cannot be tolerated, so to become successful large Glaser structures will require some kind of automated debris scavenging spacecraft that can operate on a massive scale.

Lunar-based solar power installations seem the least likely simply because the Earth-based systems can be emplaced much faster and at a lower cost. Nevertheless, I believe we will eventually see lunar solar farms.

Fusion plants will not come first, but they will eventually provide most of our energy. Once they are operating reliably, around the clock and in all weather, there will be much less need of solar and wind plants. Existing plants will not be abandoned, however, because the costs to build them will be completely amortized by that time. Another thing in their favor is that, unlike fusion, these types of energy plants will not be dependent on fuel imported from the Moon.

It is likely that the primary source of energy will change from time to time. The transition to fusion energy will be smoothed by the continued development of all other sources of energy. Because of the negative environmental and health impacts of fossil fuel pollution, there will be grass-roots political pressure to switch to less polluting technologies.

In summary, the transitions from oil to these Space-based energy systems will take decades of investment and effort, and will probably occur gradually and take several generations. As a result, there will be no sudden dislocating jolt as the changes between systems take place.

STRATEGIC IMPLICATIONS: HOPE AND PROMISE

In the short-term, every nation's strategy will remain tied to oil, and therefore each will seek to maintain a strong connection to those regions where oil is abundant. But over the long-term, perhaps beyond eighty years, there will be very little competition for oil.

Energy from Space will reduce competition for oil to below the level that inspires wars. This is what is happening now with energy – we are innovating, we are "Imagineering" as Walt Disney said, and we are going directly to Earth's primal energy source, the Sun. We are doing it in partnership with countries that were once our adversaries.

Energy independence with energy from Space, and the accompanying and necessary technological developments, will have manifold consequences. Among those are:

It will be used at central points to electrolyze water into hydrogen and oxygen, creating a much cleaner hydrogen fuel for factories, homes and vehicles. Having hydrogen available for transportation energy will eliminate the millions of automobile point sources of pollution. Where oil and coal are still burned in power plants for general electricity needs and for electricity to produce hydrogen, it will be much easier to scrub pollutants from emissions at these fixed point sources than from widely dispersed ones.

SPS and energy systems on the Moon will lessen our need to get energy from other resources, and can provide energy to any military virtually anywhere in the world.

Minerals mined in Space, such as from Near Earth Objects (asteroids and cometary cores) will be used for most in-Space construction. Because it will be relatively cheap and easy to ship cargoes from Space into Earth's gravity well, some of it will be brought to the Earth for industrial feedstock. This will seem much like "pilots of the purple twilight, dropping down with costly bales," (Alfred, Lord Tennyson, in his poem "Locksley Hall.")

Expanded access to Space will support greatly enhanced satellite platforms for communications and imagery for both civilian and military use.

The technological effort to go to the Moon and to construct solar power satellites will create a demand for frequent access to Space, thereby providing a technological advantage to the nations that pioneer these efforts.

Expanded tourist access to Space, already in its early stages, will also drive a terrific Space tourist industry, enlivening the economies around Spaceports, at large Space stations, and on the Moon.

Taken together these factors will support a thriving global economy. They will also contribute to a sense of adventure in Space, and will affect arts, literature, and music both in Space and on Earth. Many people will live and work in Space. Many others, who have not gone to Space will personally know someone who has, and the result will be that Space will no longer be seen as an unfamiliar, foreign, or hostile place, nor will it seem far away.

SUMMARY

The main theme to keep in mind is that many sources of energy are available to humans, and eventually all of them will be used. Likewise, all will be improved.

The impact each will have on human society will depend upon the shifting dynamics of government policies, business taxation, government subsidies, technology advances and technology breakthroughs. Energy sources in Space that require large Spacelift capabilities will take longer to realize, but even gradual additions to early starts will constantly improve our energy picture.

Energy from Space will gradually replace oil and gas as fuels, enhancing the environment and giving it time to recover from past damages. Space will become another workplace option for everyone. Space is a place where people from many cultures will work together in a dangerous environment. Cooperation, appreciation, tolerance and understanding between groups will be mandatory for success.

And finally, as humans move into Space we will gradually come to recognize Space as a natural part of our universal home.

•••

EDWARD B. KIKER

Edward B. Kiker attended Harvard University in the ROTC program, majoring in Lunar Geology. He worked on the Apollo 15 landing site selection. Ed served from 1971 to 1975 on active duty as an Engineer Officer at Ft. Greely, Alaska, Camp Carroll, Korea, and Fort Belvoir, Virginia. Following that, he served as an Armored Cavalry Arctic Scouts Officer from 1975 to 1979 with the Alaska Army National Guard. Mr. Kiker's civilian service and Space activity began 1975 as a Natural Resources Specialist at Fort Greely, Alaska, using LANDSAT satellite multi-spectral imagery to work with land and wildlife management. Ed served as Alaska State Director, Project High Frontier, where he developed an outline of what would become the National Ballistic Missile Defense System and commercial Space programs. He then moved to the US Army Space Institute, Fort Leavenworth, Kansas, to design the National Ballistic Missile Defense System's organization, operation and required operational capability for the Army Training and Doctrine Command. Ed represented the Army Corps of Engineers at the International Lunar Base Design Workshop in Switzerland, and presented the Lunar Mining program at the 2nd United Nations Space Conference in Vienna, Austria. Mr. Kiker also created and taught "Our Future in Space" for the University of Alaska.

CHAPTER 12

HARNESSING THE SUN:
EMBARKING ON HUMANITY'S NEXT GIANT LEAP

FENG HSU, PH.D.
NASA GODDARD SPACE FLIGHT CENTER

It has become increasingly evident that facing and solving the multiple issues concerning energy is the single most pressing problem that we face as a species. In recent years, there has been extensive debate and media coverage about alternative energy, sustainable development and global climate change; but what has been missing in mainstream media is the knowledge and point of view of scientists and engineers. This chapter elaborates on the prospects of mankind's technological capability and societal will to harness solar energy, and focuses especially on the issues of: (1) space based solar power (SBSP) development, and, (2) why it is imperative that we must embark on humanity's next giant leap - harnessing the Sun.

I. SOLAR POWER FROM A HISTORIC PERSPECTIVE OF HUMAN EVOLUTION

Solar Energy, either terrestrial based or space based solar power, has not been widely discussed in the loud and growing conversation about global warming, yet it is starting to be viewed by many scientists and visionaries as one of the most promising and feasible ways to completely overcome human dependence on fossil fuels.

At the 2007 Foundation For the Future International Energy Conference in Seattle, my presentation was one of the few that took a look back at energy use throughout human history. In this chapter, I would like to offer a brief summary of the various stages humanity has passed through in our quest for energy. In order to understand and fully appreciate the profound idea of harnessing solar energy, it is important for us to understand and learn from history, especially during crises such as the one today concerning energy and sustainability. There are no better lessons to help us foresee what may be ahead than those we can learn by looking back at the tracks of our ancestors.

There have been fundamentally three eras of energy supply and consumption in human history and prehistory. After the first fire was lit by mankind, our lives were based on wood-generated energy. We burned firewood, tree branches and the remains of crops from agricultural harvests. Starting around 1600 we found coal and entered into the 2nd era of energy uses; fossil-based energy supplies. A few hundred years later, about the middle 1800's, we incidentally discovered petroleum, and commercialized the use of oil and gas, which led to the entire modern industrial civilization. Oil and gas energy sources; being fossil-based, belong to the 2nd era. Near the middle of the 20th century, propelled rapidly by atomic energy, came the dawn of the Techno-era of energy use and production.

As the world demand for energy continues to soar, we are running into profound energy and environmental crises. These crises have been induced by the way we produce and use energy -- which predominately remains the same as the fossil-based 2nd era of human energy solutions.

Today there is great uncertainty about the world's future energy supply. If you plot the energy demand by year of human civilization on a terawatt scale, you will see a huge increase that began barely a hundred years ago. After about 150 years of burning fossil fuels, the Earth's 3 billion years' store of solar energy has been plundered. Therefore, in my view, mankind must now embark the next era of energy supply and consumption by rediscovering the mighty energy resource of our Sun. The era of taming solar energy through technology breakthroughs may well trigger the next giant leap of our civilization, elevating our species by transforming our combustion world economy into a forever sustainable Solar-electric world economy!

The Romans used flaming oil containers to destroy the Saracen fleet in 670, and, in the same century, the Japanese were digging wells with picks and shovels in search of oil, to a depth approaching 900 feet. By 1100, the Chinese had reached depths of more than 3000 feet searching for energy, all many centuries before the West sunk its first commercial oil well in 1859 in Titusville, Pennsylvania.

It is now time for humanity to look up and start digging into the sky, much like our ancestors dug into the surface of the Earth long ago. We may not achieve an entirely solar-electric civilization any time soon, but we must start now and stay on the right course – Harnessing the Sun!

II. SOLAR ENERGY:
THE ULTIMATE ANSWER TO ANTHROPOGENIC CLIMATE CHANGE

The evidence of global warming is increasingly alarming, and the urgency of a potential catastrophic climate change scenario is therefore profoundly dire. It is not certain that we can in fact overcome climate change. The place I work is Goddard Space Flight Center, a premiere NASA and world research center in the forefront for space and Earth science research, and for monitoring and analyzing any cosmic or global climate changes. I get first-hand scientific information relating to global warming issues, especially on the latest of ice cap melting dynamics or changes on both poles of our planet. I discuss these with Goddard colleagues who are world-leading experts on climate change topics. Regardless of the true or direct causes, whether it is due to human interference or cosmic cycling of our solar system dynamics, there are two basic facts which are crystal clear to all of us.

1.   There is overwhelming scientific evidence showing clear and positive correlations between the level of CO2 concentrations in the Earth's atmosphere with respect to the historical fluctuations of global temperature changes.

2.   The overwhelming majority of the world's scientific community have reached consensus that catastrophic global climate change is highly likely if we humans do nothing and continue to ignore this problem, and continue dumping huge quantities of greenhouse gases into the atmosphere.

In my view as a risk assessment expert, from a probabilistic perspective it is at least orders of magnitude more risky for humans to do nothing to curb our fossil-based energy addictions, as compared to engaging in a fundamental shift of our energy supply landscape. This is because the risks of a catastrophic anthropogenic climate change can lead to the extinction of the human species, which is a risk consequence simply too high for us to accept and doing nothing about.

In my mind, it is absurd to hear the argument made by some of our politicians that humans should not worry about "global warming" because if we do restrict the burning of fossil energies, there will be economic consequences. Whoever makes such an argument is clearly ignorant of the concept of risk, uncertainty and risk mitigation. What we are really talking about is choosing between risks. Every human activity involves risk taking, and we cannot avoid risk entirely, but only choose between them, making scientific based policy trade-offs and hopefully selecting wisely.

Therefore, there has to be a risk-based, probabilistic thought process when it comes to setting national or international policies to deal with global warming and energy issues. As the measure of Risk is a product of event Likelihood and Consequence, I believe the choice is crystal clear. When the consequence or risk of a potential human extinction due to catastrophic climate change is compared with the potential consequence or risk of loss of jobs or slowing down the economy due to restriction of fossil-based energy consumption, we must choose for the survival of humanity and accepting a much smaller risk of facing economic consequences, which are most likely short term and limited in scope, if there are any. Furthermore, by making a paradigm shift of the world's energy supply through extensive R&D and technology innovation on renewable energy production, we may well create countless new jobs and end up triggering the next spectrum of economic development and industrial revolution that mankind has ever before seen!

Hence, the acceleration and aggravation of a potential anthropogenic catastrophic global climate change is the number one risk issue of our current Combustion World Economy. Shall we continue the endless debate on the issues of global warming and sustainable human development, or would it be more intelligent to start to do something in a significant way? We have no better choices, as it is a make or break it situation.

III. SOLAR POWER:
THE BEST RENEWABLE ENERGY SOURCE FOR THE FUTURE

Humanity is now at an energy crossroads. We have two distinctive and fundamental directions to choose between.

1. Either we look for energy based on cosmic-based, open, and unlimited original resources, which means everything that comes from the stars, including our Sun, or

2. We follow the direction of using Earth-based, local and confined secondary energy resources.

For our well-being and survival, it is time for humans to tame the natural force of the Sun and harness it -- before our planet reaches the tipping point of a catastrophic climate change through continued consumption of fossil energy.

There is no doubt in my mind that harnessing the Sun is the next giant leap for mankind. Harnessing the Sun as our ancestors first harnessed fire is an inevitable and logical leap forward in the process of human evolution. Going forward, humans should learn to bypass the solar-to-fossil inefficiency. It took about 3.5 billion years and rare geologic events to sequester hydrocarbons and built up fossil deposits beneath the surface and hydrogen in the atmosphere of our planet, while using direct solar energy, could be in theory about 1,200,000,000,000 times more efficient than taming the secondary solar energy – oil, gas and coal. So why not go directly to the Sun?

The best place, of course, for a nuclear fusion reactor (which is what our Sun is) is about 149 million km away. It operates there safely and free of charge. The Sun's energy takes only 8 minutes to arrive here, leaves no radioactive waste, and is terrorist proof. It puts out about 3.8E11 TWh of energy per hour, and Earth receives about 174,000 terawatt each second. In fact, every hour Earth's surface receives more solar power than all humans use in a whole year.

IV. SOLAR ENERGY VS. OTHER FORMS OF RENEWABLE SOURCES

We must set priorities and choose wisely. Projected world energy use by fuel type in the next 30 years suggests that we are going to have an explosive increase in demand. According to recent U.S. DOE data, renewable energies including biomass, hydropower, geothermal, wind, solar, and others totaled about 6 percent of total energy production in the U.S., while nonrenewable fossil energies made up the rest. To recognize that solar energy is the best option for the future, we have to first understand our energy requirements, and decide how should we evaluate and compare our energy options for major R&D activities for the future of mankind. In my view, energy should be affordable for all human beings, inexhaustible in terms of the livable planetary lifetime, cause no harm to the environment of humans, and it must be easily available and evenly accessible around the globe. It has to be distributed in usable, flexible, and decentralized scalable forms, and there must be low risk of potential misuse for mass destruction. Energy; has to help retain and improve human values and global collaborations, must help expand human presence and survival within our solar system, and has to be consistent in elevating human culture, quality of life, and civilization.

Oil and gas fossil fuels will be depleted in another 40 to 60 years, while coal will be depleted in about 500 years, although some estimate a thousand years; but that does not really matter, because before the depletion of all coal deposits the global environment would have already suffered from catastrophic collapse. Fissionable nuclear material will be depleted in about 50 years as well.

All fossil fuels are harmful to Earth's biosphere, and nuclear power has major issues with waste deposit and the risks of proliferation and misuses. Hydro power is limited and unstable, and liquid biomass competes for land with food production. You may have heard that in Mexico tortilla prices have gone up about 60 percent in the last two years. Hydrogen (fuel cell) carries high storage and transportation risks, and it is not a source of energy but rather a form of energy storage. Wind, geothermal, and tidal sources are intermittent, unstable, and presently costly. Nuclear fusion has been studied using government funds for more than half a century, and it is unlikely to be achieved any time soon, and it has high potential for misuse.

When you carefully compare and evaluate each available option of nonrenewable and renewable energy sources against these requirements and criteria, it is evident that solar power is the most viable source of renewable energy for sustainable human development into the future.

V. THE PROSPECT OF SOLAR ENERGY DEVELOPMENT FROM SPACE

Some might ask, Why solar energy from space? Is it a technologically feasible or commercially viable human endeavor? My answer is positively and absolutely "Yes."

One of the major challenges of terrestrial solar power is the high cost of photovoltaic (PV) cells, and the inefficiency of converting the Sun light energy into electricity. Depending on the location on Earth, there is roughly 7 to 20 times less energy per square meter on Earth than in space. Based on the existing solar technology and PV materials, it would require a field of solar panels the size of the state of Vermont to power U.S. electricity needs. And, unless there are breakthroughs in the conversion efficiency of PV cells, to satisfy world demand would require about one percent of the land that is currently used for agriculture worldwide.

One of the major concepts of energy from space is Space Solar Power, SSP. SSP is a space-based energy system concept that is not a new idea, as it has been systematically studied since the middle 1970's. Long before that, Nikola Tesla, the pioneer of modern electromagnetism and inventor of wireless communication had dreamed of finding the means to broadcast electrical power without wires. Early in the 20th century, when Tesla addressed the American Institute of Electrical Engineers to explain his attempts to demonstrate long-distance wireless power transmission over the surface of the Earth, he said, "Throughout space there is energy. If static, then our hopes are in vain; if kinetic – and this we know it is for certain – then it is a mere question of time when men will succeed in attaching their machinery to the very wheel work of nature."

The SSP concept in its present form was originated in 1968, when Dr. Peter Glaser first developed the idea of SSP as a source for continuous power generation for the Earth's future energy needs. Glaser's basic idea was that satellites in geosynchronous orbit would collect energy from the Sun, and the energy would be converted to radio waves and beamed to receiving sites on the ground. The ground antenna would then reconvert the radio waves to electricity.

In our current, more refined version of the SSP system, solar energy is collected in space by satellites in a geostationary orbit. It is then converted to direct current by solar cells, which power microwave generators in the gigahertz frequency range. The generators feed a highly directive satellite-borne antenna, which beams the energy to the Earth. On the ground, a rectifying antenna (rectenna) converts the microwave energy to direct current, which, after suitable processing, is fed into the terrestrial power grid. A typical Solar Power Satellite unit, with a solar panel area of about 10 square km, a transmitting antenna of about 2 km in diameter, and a rectenna about 4 km in diameter – may yield an electric power of about 1 GW, the equivalent of a large scale nuclear power station.

Among the key technologies involved in SSP are microwave generation and transmission techniques, wave propagation, antennas, measurement, beam control and calibration techniques. Key issues include potential effects on humans and the potential interference with communications, remote sensing, and radio-astronomy observations.

VI. THE TECHNOLOGICAL AND COMMERCIAL VIABILITY OF SPACE SOLAR POWER

Is SSP a viable option? In my opinion, it can be a viable energy option for base-load electricity generation to power the needs of our future. SSP satisfies every major criterion of a viable energy option, with the exception of the cost based on current space launch and propulsion technology: space transportation cost is one of the major hurtles for SSP.

To overcome the high launch cost, the development of a Reusable Launch Vehicle (RLV) and autonomous robotic technology for in-orbit assembly of large solar structures is needed, along with systems to assure safety and reliability for these large and complex orbital structures. Nevertheless, there are no breakthrough technologies that need to be invented, nor any theoretical obstacles that need to be overcome for an SSP project to be carried out.

The U.S. government provided about $20 million to study SSP in the late 1970's, but then abandoned this project with nearly zero dollars spent up to the present day. Today, a government funded SSP demonstration project is absolutely long overdue. The excellent book *Sun Power: The Global Solution for the Coming Energy Crisis* by my friend, Ralph Nansen, offers a detailed history on the subject. Ralph was the Boeing manager of the DOE-NASA funded SSP proof of concept study in the late 1970's, and published *Sun Power* in 1995, accurately predicting our current situation. Dr. Peter Glaser's book *Solar Power Satellites: A Space Energy System for Earth* also offers superb reading on this topic.

We can solve the cost issue to make SSP a commercially viable energy option through human creativity and innovation on both technological and economic fronts. Besides the continuing quest for a low-cost reusable launch vehicle (RLV), there are other possibilities for ingenious commercial or business models that could overcome the SSP cost issues.

One model now being pursued by the Space Island Group, an American private aerospace entrepreneurial company based in California, is to use modified Space Shuttles by turning the huge volumes of the external tanks into commercial assets for space-based research and orbital tourism. A huge demand in space tourism will certainly bring about a higher launch rate, and that will in-turn drive down the space transportation cost, thereby helping to make SSP more viable. If we compare this with the commercial aviation industry, who would have thought that ordinary people could afford air travel just several decades after Wright Brothers had succeeded in their first aircraft test?

Further, we do not need to restrict our vision to choosing between terrestrial solar and SSP. In fact, the dream of SSP can be realized much sooner through advocating the use of terrestrial solar energy and engaging in the pertinent R&Ds on a grand global scale. This is because the advancement in major terrestrial solar technologies such as nano-particle based ultra high efficiency and low weight, low cost PV cells, along with super capacity and low cost energy storage systems will also support affordable terrestrial and SSP development.

Our ultimate goal is to tame the "very wheelworks of nature" and harness the Sun. With rapid advances in nanotech-based PV solar cell material, now reaching over 50% efficiency, and which can be cheaply produced (along with revolutionary battery technologies), it is totally possible that one day we won't have to launch huge PV structures into

Earth orbit to satisfy the base-load electricity consumption requirements of the entire planet.

It's extremely exciting to see in recent years, the rapid advances in the PV cells research, and an over 30% annual growth of solar energy production, even without government policy support from major countries such as the U.S. and Russia. If every house in the future were built with cheap and highly efficient solar cell materials on the roofs and sidings, and every shaded parking lot in shopping malls and office buildings was built and equipped with solar powered charging plugs for electric cars, then how different our energy picture would be!

VII. ACHIEVING ENERGY FROM SPACE: A ROADMAP AHEAD

The realistic hope of a commercially viable SSP system lies in a collaborative effort between the emerging private, entrepreneurial space businesses and venture capital investment. I am not optimistic about government involvement in this great human engineering and technological endeavor, especially on the part of the much needed support from the U.S. government. But I am happy to see that great private sector visionaries see the significance of future energy systems as part of the vision for space exploration.

One such visionary is the recently retired president of India, Dr. Kalam Abdul. Dr. Abdul had the great courage to speak publicly on SSP while addressing the Symposium on "The Future of Space Exploration" organized by Boston University in January 2007. Dr. Abdul noted that space research is truly inter-disciplinary, and has enabled true innovations at the intersection of multiple areas of science and engineering. He also noted that, "Civilization will run out of fossil fuels in this century. However, solar energy is clean and inexhaustible. And while solar flux on Earth is available for just 6-8 hours every day, incident radiation on a space solar power station would be 24 hours every day. What better vision can there be for the future of space exploration, than participating in a global mission for perennial supply of renewable energy from space"?

Government support for policies and financial resources for R&D and the related technology demonstrations are crucial to the success of such giant effort. There was no U.S. government sponsored work until NASA initiated their "A Fresh Look" studies in the mid 1990's. Subsequently, however, the Department of Energy abstained from involvement.

However, during this period the Japanese government and industry became interested in the concept, and updated the reference system design developed in the System Definition Studies in the late 1970's, conducted some limited testing, and proposed a low orbit 10 megawatt demonstration satellite. Their effort has been curtailed by their economic problems and by their lack of manned space capability. SSP Interest by other nations has persisted however, but only at low levels.

The overwhelming initial cost of development and deployment has remained the primary obstacle. As I noted, number one on the list of cost barriers is the cost of transportation: solar power satellites are only economically feasible if there is low cost space transportation. Therefore, in order for SSP to be successful, we need an organized consortium consisting of private businesses, venture capitalists from major international partners, along with government support of major industrial nations on R&D and technology demonstration. We need this in order to bring down associated project and technology risks concerning safety, reliability, and technology maturity. A consortium-based Comsat model as was used for successful launch and commercialization of communications by the satellite industry should be a viable approach for SSP.

VIII. AN APOLLO-LIKE PROJECT OF SPACE SOLAR POWER

A major Apollo-like effort, led by the U.S. with participation from the broad international community, may be what is needed to successfully create, implement and operate a commercial-scale SSP system.

An inherent feature of solar power satellites is their location in Earth orbit, outside the borders of any individual nation, with their energy delivered back to the Earth by way of wireless power transmission (WPT). The applications of WPT must be compatible with other uses of the radio frequency spectrum in the affected orbital space. Therefore, in order to prevent international confrontation, it is vital for there to be international governmental involvement in coordinating global treaties and agreements covering frequency assignments, satellite locations, space traffic control and many other aspects of space operations.

I believe it is imperative for a multi-governmental organization or entity to be put in place for a major SSP project, because it would be extremely difficult, if not inconceivable, for the U.S. or any other single nation to do this alone at any useful or significant power scale, due to the many political and technological issues. However, it is equally important that there must be a nation to provide the necessary leadership in the complex and interdependent international SSP effort. To prevent chaos in a partnership of multiple governments and industries, it is vital that the leadership and responsibilities of the various project elements be clearly defined. The United States is the logical leader in this area -- because of the breadth of technology infrastructure and capability that already exists, as well as the magnitude of financial resources available within its industrial and financial communities.

Space solar power is going to be a gigantic, achievable human technology and engineering endeavor. Mankind had achieved going to the Moon and splitting atoms nearly half a century ago; we can certainly overcome the inefficiency problems of the solar-electric conversion, and we can achieve the goal for affordable access to space, hence making SSP a cost competitive energy source for all of humanity.

Key SSP component technologies will also enable human economic expansion and settlement into space, which may also be important for the permanent survival of our species. To this end, such a "vertical expansion of humanity" into our solar system in the new millennium can be every bit as important (if not far more critical) as the "horizontal expansion" that was achieved by our ancestors.

Indeed, SSP will provide an ideal platform for promoting human collaborations, and in-turn help reducing global economic imbalance. It can be also a major steppingstone for humanity's next giant leap forward by harnessing the Sun and transforming the combustion world economy into a solar-electric human civilization.

IX. LOOKING FORWARD TO AN EVER BRIGHT FUTURE

It is time for humanity to look to the Sun for answers to our ever increasing energy needs, and to solve our environmental and economic fossil fuel crises.

I suggest that "harnessing of the Sun" will be the 3rd giant leap in the process of human evolution. The first was when human beings got down from the trees and started to use fire, which led to tool-making, agriculture and ancient civilization. Then humans invented machinery and discovered electricity, which allowed us the 2nd giant leap forward to modern industrialization. Now humans are running into profound energy and environmental crises, and we must now embark on the next giant leap of civilization. By harnessing the Sun to transform combustion civilization, which was built upon primitive and secondary energy sources, we can build our way and evolving into a solar-electric civilization, fueled by the inexhaustible and direct energy source from the stars.

Can mankind achieve the third giant leap into the solar-electric civilization? My answer is positively YES! Humans are capable of profound achievements; including the huge Manhattan and Apollo projects. We can certainly succeed in taming the mighty power of our star! In my view, the key changes and support needed for SSP are much less technical or economic than social and political; however, together we can make it happen by educating and mobilizing politicians and decision-makers around the globe.

Indeed, it is a policy issue more than a technology or economic issue. As Dr. Robert Goddard liked to say, "It is difficult to say what is impossible, for the dream of yesterday is the hope of today and the reality of tomorrow." As I noted at the end of my talk at the Seattle energy conference, "As intelligent creatures rooted in the cosmic origin, humanity was meant to survive and spread its presence all over the universe by milking the energy of the stars!"

•••

FENG HSU, PH.D.

Dr. Feng Hsu is a well respected world expert on risk and risk assessment. He is a former research fellow of Brookhaven National Laboratory in the fields of risk assessment, risk-based decision making, safety & reliability and mission assurances for nuclear power, space launch, energy infrastructure and other high integrity social and engineering systems.

He currently heads the NASA Goddard Space Flight Center risk management function, and is the GSFC lead on the NASA-MIT joint project for risk-informed decision-making support on key NASA programs, such as the GPM (Global Preciptation Measurement), LSS (Lunar Surface Systems) and the CxP (Constellation) etc. He was also a leading engineer/scientist in the Shuttle and Exploration Analysis Department at Johnson Space Center, SLEP and Shuttle upgrade trade studies etc.

Dr. Hsu serves on many agency and center expert panels supporting challenging SMA issues. He is co-chair of several international technical committees. He played key roles in the: (1) STS-107 (Space Shuttle Columbia) investigation team, (2) the RTF (Return to Flight) team, and (3) the ECO (Engine Cut-Off) expert team for the Discovery mission. Dr. Hsu has over 90 publications and is co-author of two books. He is frequently invited to be the keynote speaker in many international forums. His recent interests span from challenges to human space exploration to solar energy; and include global collective intelligence and risk based policy-making on emerging environmental and energy security issues.

Dr. Hsu holds a Bachelors degree in Applied Math, a Masters degree in Operations Research and Statistics and a Doctoral degree in Engineering Science. As a senior advisory member of the Aerospace Technology Working Group (ATWG) and a co-founder of Space Development Steering Committee, Dr. Hsu has been a strong advocate for Space Based Solar Power (SBSP) for years, and was instrumental in instigating the 2007 NSSO study on SBSP. Dr. Hsu has contributed whole-heartedly to the great human endeavor of harnessing solar energy for sustainable human development.

CHAPTER 13

ANTIMICROBIALS AND LONG-DURATION SPACE FLIGHT

JEROME BELL, B.S.
BIODRI TECHNOLOGIES, LLC

1. INTRODUCTION

Previous research studies and flight experiments conducted by NASA have shown that microorganisms thrive in space, and therefore they present a potential threat to humans, flight systems, and to the environments of other planets and of Earth.[1] The presence of Earth-originated microorganisms on spacecraft can also affect other aspects of spacecraft and habitat operations and scientific results, such as observations conducted on other planets for signs of life, human hygiene accommodations, and even the quantity of clothing that needs to be stowed on board.

For long-duration space activities, the evidence strongly suggests that systems that control microorganisms are mandatory. Control needs to focus on three areas: (a) detection and identification of microorganisms; (b) prevention of the formation and growth of microorganisms; and (c)

---

[1] Christensen, Bill, "Unwanted Life Forms Abound in Sick Spacecraft," Technovplqycom, posted 22 May, 2007.

treatment or mitigation of the effects produced by unwelcome microorganisms.

For example, the use of antimicrobials in crew apparel has drawn the attention and interest of the NASA Johnson Space Center, since it would provide a means of reducing the quantity of clothing that astronauts would be required to take on long-duration space missions to the Moon or Mars.

"Among the many issues affecting long-duration human space activities will be the need to control the growth and spread of microorganisms in a closed environment. Crew members and their clothing, bedding, and other fabrics with which they come into contact are potent sources of such microbial contamination. These microorganisms could be suppressed by frequent disposal or frequent washing of the affected fabrics, though neither solution is practical in a weight-limited nor a water-limited closed environment."[2]

As we look to the future, all space activities, including those undertaken by the U.S. government, private space operators, other nations, or multi-nation partnerships will need to provide effective defense against microorganisms in all phases of space operations: prelaunch, during flight, on other planets, and upon return to Earth.

For these reasons, a variety of facilities will be required to further develop and test the private and public sector's science and technologies. This development and testing should include both the Earth and outer Space environments.

This chapter evaluates current commercially available antimicrobials, and discusses the strengths and weaknesses of each.

2. TYPES AND CHARACTERISTICS OF MICROORGANISMS

While many microorganisms are beneficial to humans, many others are pathogenic. Pathogens fall into four main categories: bacteria, viruses, fungi, and protozoa.

BACTERIA
Perhaps the most familiar "germ," bacteria are single-celled organisms. They can reproduce both inside and outside the human body, and they can do so extremely quickly. One bacterium could become one billion in just 10 hours. As a result, bacteria can be found living on almost every surface in every climate on Earth. A single teaspoon of soil can contain more than one billion.

---

[2] NonReimbursable Space Act Agreement Between NASA Johnson Space Center and BioDRI Technologies, LLC for Development of Improved Fabrics for Space Flight for Clothing and Cabin Interiors, SAA-AT-06-002, February 22, 2006.

Viruses

Viruses are simpler organisms in structure than bacteria, being the smallest of the microbes. However they possess the same ability to clone themselves and reproduce exponentially. Unlike bacteria, viruses need to be in or on a living host (plant, animal, or human) to grow and reproduce. Viruses travel via air currents or through body fluids, and are constantly looking for hosts to invade. They are able to survive on surfaces for days, and only a few are needed to be able to infect.

An invaded cell loses its ability to function normally, and is forced to follow the viral instructions to produce viral proteins which create more clones of the invading virus. Once these clones are assembled, the virus forces the host cell to rupture, thereby releasing the clones which then infect other host cells, and thus disease spreads.

Fungi

Fungi include yeasts, mushrooms, and molds. Unlike other plants, fungi don't draw nourishment from soil, water, or air. Instead, they feed on plants, people, and animals, and they thrive in damp, warm places and in poorly lit, poorly ventilated locations. They can cause irritation and disease as they colonize substrates (including your skin), and release spores through asexual reproduction.

The biggest problem with fungi is that they are very difficult to kill without also harming the host cells. Thus, drug treatments tend to focus on preventing further growth; but do not manage to kill off existing colonies.

Protozoa

These are one-celled organisms, such as amoebas, that often spread through moisture and water, and cause a range of diseases from gastro-intestinal infections which lead to diarrhea, nausea, vomiting and stomachaches, to other serious illnesses, such as dysentery and malaria. Some protozoa are parasites and must live in other organisms.

Regardless of the type, these "germs" are usually spread through contact via perspiration, blood, saliva, and through the air via sneezing, coughing, and even breathing.

Simple cleaning and good hygiene practices can limit their growth, and preventing grime and dust collection helps to eliminate the environments in which they thrive.

3. How Antimicrobials Work

Microbes feed, breed, and thrive in environments that are referred to as 'biofilm.' The basic objective of an antimicrobial agent is to kill microorganisms, and to neutralize or delay reformation by eliminating the biofilm.

A bacterium cell structure has three major elements: internal structure, surface structure, and appendages.

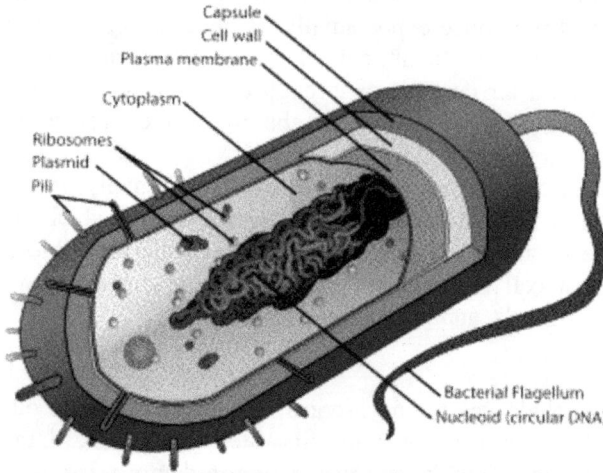

Source
http://commons.wikimedia.org/wiki/Image:Average_prokaryote_cell-_en.svg

- The Internal Structure is the nucleoid that contains the DNA.
- The Outer Structure consists of the capsule that protects the bacterial cell and the cell wall that maintains overall shape. There are 3 primary shapes, spherical, rod-shape, and spiral.
- There are numerous proteins moving within or upon the plasma membrane layer. They are primarily responsible for transport of ions, nutrients, and waste across the membrane.
- Appendages include pili, hollow, hair-like structures made of protein that enable bacteria to attach to other cells, and Flagella, which provide cells the ability to move spontaneously and actively.

Unlike antibiotics, which target specific microorganisms residing inside a human or animal, antimicrobial treatments are generally indiscriminate as to the type of microorganism targeted, and are therefore suited only for external use, since they will kill beneficial and necessary microorganisms as well as pathogens. Some antimicrobial treatment products fall under the jurisdiction of US regulatory agencies and regulations, including the EPA and FDA, and their use in space may require more stringent limitations. Waivers may be need to be granted to address unique or specialized requirements.

Microorganisms need a host on which to live and eat, and it is the biofilms that provide that host. Biofilms are populations or communities of microorganisms that adhere onto environmental surfaces. These microorganisms are usually encased in an extracellular polysaccharide that they themselves synthesize.

Biofilms may be found on any surface where sufficient moisture is present: including dust particles in the air that contain skin cells, sneezes and coughs, animal hair and dander, pathogens, yeast, mold spores, pollen, smoke from various sources, and moisture molecules. They can also be found on surface liquids, in blood, in kitchens and in bathrooms.

They develop rapidly in flowing systems where nutrients are available. Bacteria living in a biofilm usually have significantly different properties from free-floating bacteria of the same species; as the dense and protected biofilm environment allows them to cooperate and interact in various ways. One effect of biofilm is to increase the resistance of bacteria to detergents and antibiotics because the dense extracellular matrix and the outer layer of cells in the biofilm protect the interior of the community.[3] Therefore, if biofilm-hosting surfaces are simply sprayed and the microbes are not completely removed, the net quantity of pathogens can return to the original state within as little as 10 to 15 minutes.

Biofilm is formed in phases, as illustrated below.

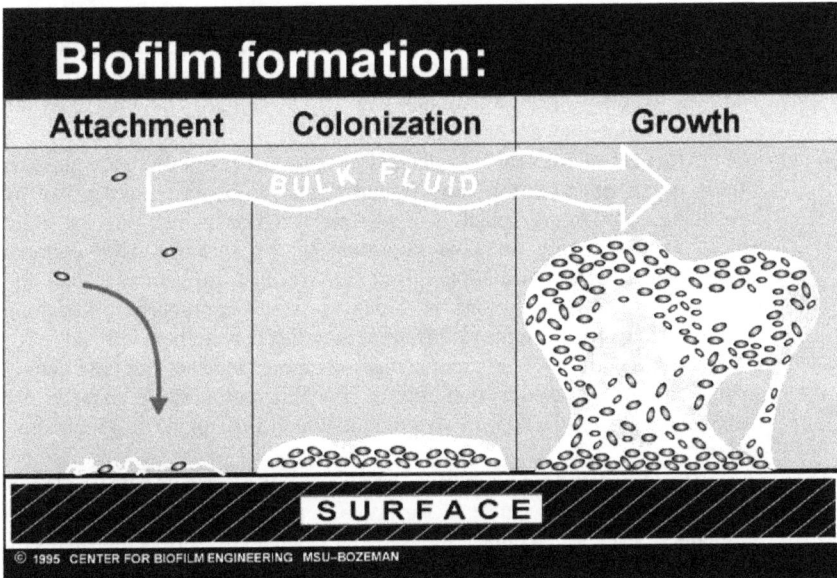

Printed with the permission of "MSU Center for Biofilm Engineering, P. Dirckx".

---

3   Stewart P, Costerton J (2001). "Antibiotic resistance of bacteria in biofilms." *Lancet* 358 (9276): 135–8. doi:10.1016/S0140-6736(01)05321-1. PMID 11463434

The bulk fluid shown in the figure could be air passing over a surface, a stream in open air, or a fluid passing through a pipeline.

The following section is adapted from: "A Comparison of Antimicrobials for the Textile Industry," W. Curtis White M.S., James W. Krueger, and Robert A. Monticello Ph.D., ÆGIS Environments, Midland, MI, USA, and Patrice Vandendaele, Devan Chemicals, Renaix, Belgium.[4]

There are two different classes of antimicrobials. In either case, the wall of a living cell must be penetrated. One class is referred to as unbound (commonly known as leaching) and the other class as bound. The distinction is whether the antimicrobial molecularly bonds to the surface it is applied to. Leaching antimicrobials kill by way of poisoning or chemically altering the organism, while Bound antimicrobials kill by way of bonding to a surface, attracting organisms through difference in electrical charge, spearing the organism and subsequently killing by 'electrocution.'

UNBOUND CLASS ANTIMICROBIALS

An unbound antimicrobial must diffuse or leach from the treated surface and be consumed by a microorganism. Once inside, the chemical agent acts as a poison, interrupting key metabolic or life sustaining process of the cell, causing it to die. Once the antimicrobial is depleted or washed away during regular maintenance, protection vanishes. After application, an unbound antimicrobial continues to diffuse or leach from the treated surface. The vast majority of antimicrobials work by leaching or moving from the surface on which they are applied.

Most antimicrobials are intended to act quickly and dissipate quickly. Others use a time release capsule and obtain a longer working life by imbedding the antimicrobial in a material such as paint, glue, or other coating, and counting on slow migration to the surface. Conventional antimicrobials, even those applied in a carrier, must diffuse (wash off) and create a "zone of inhibition" in order to function properly. Leaching antimicrobials include compositions that may incorporate heavy metals such as arsenic, lead, tin, mercury, and copper and these elements can pose danger to humans, animals, and treated objects. For this reason, these elements are increasingly being placed on Government Watch List for further regulatory action.

Besides affecting durability and useful life, leaching technologies have the potential to cause a variety of other problems when used in garments. They can contact the skin and potentially effect normal skin bacteria, cross the skin barrier, and/or have the potential to cause rashes and other skin irritations.

---

[4] *A Comparison of Antimicrobials for the Textile Industry*, W. Curtis White M.S., ÆGIS Environments, Midland, MI, USA, Robert A. Monticello Ph.D., ÆGIS Environments, Midland, MI, USA, James W. Krueger, ÆGIS Environments, Midland, MI, USA, Patrice Vandendaele, Devan Chemicals, Renaix, Belgium Copyright 2000 AEGIS Environments.

BOUND CLASS ANTIMICROBIALS

A different type of antimicrobial is molecularly bonded (bound), which relies on the technology remaining affixed to the substrate, killing microorganisms as they contact the surface to which it is applied. When applied, these antimicrobials actually polymerize with the substrate, making the surface itself an antimicrobial agent.

4. SUMMARY OF SPACE ACT AGREEMENT RESULTS

NASA has recognized the need to control the growth and spread of microorganisms in a closed environment for long-duration human spaceflight missions – where crew members and their clothing, bedding, and other fabrics, as well as the surfaces on the interior of the spacecraft itself, are all potent sources of such potential contamination.

The author, along with other principals of the BioDri Corporation, has conducted collaborative research with NASA to begin to evaluate the capacity of several different kinds of antimicrobial agents on various textiles to resist long-term microbial growth.

(A) ANTIMICROBIAL GLOVE ASSESSMENT

At the request of the Astronaut Office, a glove with antimicrobial that could be worn under the EVA glove was delivered for Astronaut evaluation. The goal was to reduce or prevent itching and perspiration on astronaut's hands caused by the outer glove during EVA. The test was successful.

(B) WEAR TEST OF ANTIMICROBIAL SOCKS

Eleven NASA personnel were provided with antimicrobial socks containing Copper Oxide. There were no restrictions on the use or care of the sock. On a scale of 1-10 where 1 is worst and 10 is best, five astronauts rated the sock 10, two rated it 9, two rated it 8, one rated it 7, and one rated it 5.

(C) OUT GASSING TEST

A sample of nylon/spandex fabric with antimicrobial copper treatment was tested at JSC's White Sands Test Facility in January 2007.[5] The tests were conducted at 14.7 psi atmospheric pressure with 75% Nitrogen and 25% Oxygen. The test included an Upward Flame Propagation test and a Determination of Outgassed Products Test. While the material sample passed established NASA criteria, further testing is

---

[5] Materials Test Data Transmittal, Letter form Harold Beeson, Ph.D., Chief Laboratories Office, White Sands Test Facility to Jacobs Sverdrup Attn: Ms. Christina Hardeway, February 22, 2007.

required to determine its performance under other environmental conditions, such as lower pressure, etc.

(D) LIST OF FLIGHT ITEMS THAT COULD POTENTIALLY BENEFIT FROM ANTIMICROBIAL MATERIAL OR TEXTILE

NASA provided the following list of items that should be considered candidates for new advanced materials and antimicrobial treatments. The list includes both Space Suit/EVA systems and CEV (Orion) and portable equipment.

EVA /Space Suit items that could benefit from antimicrobial material or textile:

Textiles: Socks worn inside the suit, Thermal comfort underwear worn inside the suit (optional for crew), Maximum Absorption Garment (MAG) (diaper) liner, Communications Carrier Assembly (CCA) (Snoopy Cap), Liquid Cooling and Ventilation Garment (LCVG), Comfort gloves (worn inside of space suit gloves), Headband.

Solid surfaces applications: Inside of helmet, Inside of suit and gloves (some surfaces are soft and some are hard), CCA Microphone, Valsalva device, In-suit Drink Bag (IDB).

CEV (Orion) Habitation and Portable Equipment items that could benefit from an antimicrobial material or textile:

Textiles: Stowage Bags and/or material to prototype stowage bags, Straps for tie down, nets, and webbing materials, Towels for crew (small, similar to Russian wetted towels), Sleep Restraint liner, Clothing (socks, exercise clothing, shirt, shorts, etc.), and Wipes for cleaning.

Solid Surface Applications: Coating or Spray for Waste Management System or hygiene areas.

Other: Trash Bag liner to contain wet and dry trash, odor proof, able to be sealed to keep liquids in, anti-microbial

5.   REGULATORY AND TECHNICAL ISSUES AND CHALLENGES

5.1 REGULATORY ISSUES AND CHALLENGES

Any substance or mixture of substances that acts against microbes (such as a disinfectant, sanitizer, or sterilant) is an antimicrobial agent. Antimicrobial products for hard surfaces should not be confused with agricultural pesticides that are used to treat crop diseases, nor with antibiotics used to treat human and animal infections.

Antimicrobial products are regulated by one or more of three government agencies including the EPA, FDA, or USDA, depending on their intended use.

Commercial products are subjected to extensive efficacy and toxicity testing before they can be sold, and prior to marketing an antimicrobial product, the manufacturer must obtain EPA registration by submitting the product label and extensive data on chemistry, toxicology, and efficacy (effectiveness)

The entire composition of an antimicrobial including the active ingredients, the inert ingredients, and the formulation (the mixture) are regulated. The active ingredients are typically small quantities of biocidal components that control the microorganisms. The inert ingredients are the additives to make the product more functional: such as safer, easier to handle, measure, apply etc.

In order to obtain EPA registration, sterilants, disinfectants and sanitizers must pass a series of rigorous efficacy tests to verify their manufacturers' claims. Sterilants are subjected to the highest efficacy standards, followed by disinfectants, and then sanitizers. Toxicology screening and thorough chemical characterization are also required by EPA for each antimicrobial active ingredient and inert ingredient.

It is important to be aware that the resistance that has developed in microorganisms due to excessive use of antibiotics has not been observed to occur in microorganisms as a consequence of their treatment with antimicrobial cleaners or disinfectants, and therefore the same disinfectants continue to work even on drug-resistant strains.

More specific information about antimicrobial regulations and requirements can be found at the National Pesticide Information Center web site and the antimicrobial link that links to the EPA web site: http://npic.orst.edu/reg2.htm#antimic.

5.2 REGULATORY ISSUES AND CHALLENGES FOR SPACE MISSIONS

There are important questions to be addressed concerning the role of the regulatory agencies that govern the development and use of antimicrobials in future space missions. These would be requirements and approvals in addition to those imposed by NASA, DOD and private/commercial space operators for safety, performance, etc.

The questions below will need to be addressed by a consortium of the governmental policy setting and regulatory agencies, program implementing or controlling entities that set or are affected by the associated policies, delegated others, and finally mission requirement planners.

(a) To what extent, if any, would these regulations and agencies be applicable for development of an antimicrobial intended to be primarily used in space and on other planets, especially on the return to Earth?

(b) What additions or changes to the regulatory process would be required for using an existing commercial antimicrobial in space, or a space-developed antimicrobial for commercial use on Earth?

(c) Does the role of Earth's regulatory agencies still apply if the product is developed only for space use and only used in space?

(d) Would commercial space operators be required to comply, and would NASA and DOD be exempt from these regulations?

(e) Would it be necessary to prepare an environmental impact statement for space missions returning to Earth that use antimicrobials that are not subject to the approval process?

The ultimate answers to these questions will have a direct bearing on establishing the viability of existing commercial antimicrobials, establishing the need for further product improvements, required product technology development, compliance regulations on the part of Space Operators, specific program requirements on design and operations, and all the required testing and evaluations.

5.3 TECHNICAL ISSUES AND CHALLENGES FOR LONG-DURATION MISSIONS

There are some interesting issues and challenges that need to be addressed relative to microorganism control on long-duration missions.

A. DEVELOPMENT OF A PERMANENT ANTIMICROBIAL

What should be considered as a permanent antimicrobial? And how is it tested for permanency? One definition may be based on the life of the microbe that application/product it is applied to. Rather than setting an arbitrary time for which an antimicrobial must remain effective, it is probably more practical to base the time requirement on the useful life of the article the antimicrobial is applied to. Another might be based on the duration of uninterrupted effectiveness on a mission in the event the product lifetime is extremely long. This becomes a question of verification, certification procedure, and application methodology that requires test and demonstration in both ground and space environments.

B. DEVELOPMENT OF ANTIMICROBIALS USING NANOTECHNOLOGY

Antimicrobial agents that are biodegradable, and therefore environmentally friendly, need to be firmly bonded to a surface to be of permanent or significant-duration effectiveness. It is believed that nanotechnology offers an opportunity to blend the two attributes; but the specific applications need to be developed.

C. OUTGAS AND MICROBIOLOGICAL ACTIVITY TESTING

Because commercial antimicrobials are developed and approved for use under Earth atmosphere and gravity conditions, antimicrobial materials and textiles need to be tested for outgassing and microorganism activity under the flight environments they will be used. To ensure maximum safety, all antimicrobial agents need to be tested in their respective operating environments. This will, unless the efficacy of the agents is proven beforehand, create a situation of heightened vulnerability for the initial builders and inhabitants on the Moon and on Mars.

D. MICROORGANISM STRAIN TEST SPECIMEN

An experiment conducted on Shuttle Flight STS -115, launched September 9, 2006, revealed that a Salmonella typhimurium microorganism became three times more virulent than on the Earth, possibly due to hard-to-control biofilms. It has been noted that, "bacteria express different sets of genes in different environments to ensure their survival. Inhospitable conditions, for example, can turn on a 'master switch' in some bacteria and allow the microbes to form tough spores that can survive the extreme conditions of space."[6]

Is there an existing commercial antimicrobial product adequate for microorganisms in space? If there is not, several key questions are raised.

Will a modification in the formulation be required? Will a totally new antimicrobial need to be developed? Will testing of an antimicrobial for use in space require a space-exposed microorganism specimen? Will a space environment be required for the testing, and if so, can it be performed in an environmental facility on Earth; or will the development and testing need to be accomplished in space? The answers to all these questions can be determined through testing, evaluation, and consideration of specific mission requirements.

E. PLANETARY PROTECTION

NASA's microbial pathogen protection activities are focused on two objectives: (1) measures to protect extraterrestrial bodies from infection by microorganisms originating from Earth, and (2) measures to protect Earth from microorganism sources originating from extraterrestrial bodies that are brought back to Earth. Considering the revelation that the Salmonella typhimurium microorganism became three times more virulent in space than on the Earth, a third focus is also a worthy candidate for inclusion; that being, (3) measures to protect Earth from microorganisms originating from

---

[6] Mosher, Dave. "Space Makes Bacteria More Dangerous," *LiveScience*, 24 September 2007.

Earth and returning after a prolonged duration in space and/or destinations significantly different from low Earth orbit.

Planetary protection is not exclusively about pathogens or infections. On extra-terrestrial bodies, it is also about maintaining a pristine environment that will not comprise science or inhibit the search for life. Another goal is to prevent the spread of pathogens that are not of Earth origin.

In our efforts to reduce all types of contamination, in the past spacecraft intended for one way missions to extra-terrestrial bodies have been subjected to strict quarantine and clean room control prior to launch. For human missions, contact with crews prior to launch is restricted, and specific constraints are imposed upon those who come in contact with the spacecraft.   During Apollo, a specially-built quarantine facility was constructed to house the astronauts upon their return from the moon. The astronauts remained in quarantine for 14 days; which was deemed enough time to ensure they had not brought back any E.T.s in the form of microorganisms. After Apollo 14, this was no longer performed because no sign of E.T. life was detected.

It would be prudent to take the same precautions with and samples and crew returning from Mars or any other planet or space mission. The procedures in place do not specifically address the Earth-originated round trip microorganism that has been exposed to the space/extra-terrestrial environment over a period of time and has had an opportunity to mutate. Even though no microorganisms of lunar origin have been discovered, that does not necessarily mean that microbes brought to lunar habitats and subjected to long-duration space exposure wouldn't necessitate the same planetary protection procedures as the initial Apollo lunar missions.

5.4   POTENTIAL CONSEQUENCES OF UNCONTROLLED OR INADEQUATE CONTROL OF MICROORGANISMS DURING A LONG-DURATION SPACEFLIGHT

Space experience, space experiments, and discoveries on Earth have shown that microorganisms have the ability to exist and survive in extreme environments. For instance, mold was found to be growing in the Russian MIR Space Station. Bacteria have been found to become more virulent after short duration space Shuttle missions. Despite our attempts to control bacteria in clean rooms on Earth, they are sometimes present.

In the absence of adequate control measures, a few risks that surviving and/or striving micro-organisms may impose will be to endanger crew health, air quality, spacecraft system performance and/or operations, food and water integrity, interference with accomplishment of mission objectives, and compromising the results of scientific experiments. Additional provisions will be required to be available to respond to any outbreak.

Clearly, microbes must be dealt with on three separate levels to mitigate or significantly reduce their detrimental effects.

Devices that can quickly detect and identify the presence and types of microorganisms onboard and in real time (automatically if possible) provide early notification that action should commence. Without this, crew time would be required for routine cleaning and wiping, reducing crew availability for other activities.

Antimicrobial agents that will both kill and then prevent or significantly restrain future and/or further growth of microorganisms will be required. Preferably, these antimicrobials will be permanent and can be embedded in clothing, crew equipment, space craft structure, and equipment including air filters. Antimicrobial products for crew hygiene such as handwashes, etc., will be needed, as well as sprays for food to prevent bacteria and spoilage, especially where crops are grown in space.

Finally, since at this time no strategy can be assured to be foolproof for an undefined duration or microorganism, there must be provisions to deal with a residual presence (at a minimum) of these microbes that crew members may be exposed to. Medicines for treatment or prevention of human illnesses need to be available, especially if the microorganisms have the potential to threaten human immune systems.

6. Conclusions And Recommendations

To date, control of microorganisms in space on long-duration missions has primarily focused on research and technology leading towards equipment for detection, identification, and measurement of microorganisms in situ and real time, and research in the behavior and characterization of microorganisms in space for the objective of development of antibiotics for the crew. It does not appear that NASA or any other space organization has given significant effort towards the research, development and/or use of antimicrobials as a component for controlling microorganisms in space.

Control of microorganisms and the effects they create cannot be satisfied by either antimicrobials alone or by antibiotic medicines for humans; it's going to take both types to deal with all possible consequences of microorganisms in space. A recommended approach (philosophy) toward dealing with this issue would be expressed as a strategy to "prevent, and protect whatever cannot be prevented." Antimicrobials would be the products for prevention of microbes, and both antibiotics (for humans) and antimicrobial / antibacterial treatments (for non-human surfaces) for protection against the effects of the microorganisms. This approach therefore suggests that research and development for controlling microorganisms must address all three components.

   The remaining Shuttle flights and the continuing work on the
International Space Station both provide excellent opportunities to conduct
rigorous testing of commercial antimicrobial treatments, materials, and
fabrics in order to determine which products and methods will be most
effective on long-duration flights beyond low-Earth orbit, where the
options for external support or intervention are negligible.

                                    •••

JEROME (JERRY) BELL, B.S.

 Jerome (Jerry) Bell, a native Houstonian, earned his Bachelor
of Science Degree in Aerospace Engineering from the
University of Texas, Austin in 1963. Following 9 months of
Employment with the McDonnell Aircraft Company in St.
Louis working in aerodynamics supporting the Phantom 2
Aircraft Program, he accepted a position with NASA Manned
Space Craft Center (now Johnson Space Center) where he
worked as a civil servant for over 42 years until retirement in
2006.
   He has worked on every manned space program from Project Gemini in
various technical and Program Management capacities and Program Phases
including Program formulation, requirement and concept development, operations,
and technology in selected areas. Jerry has served as a JSC representative to
Several Agency wide activities including the NASA Space Station Task Force
Concept Development Group at NASA HQ, Non Advocacy review teams, Proposal
review and Source Board technical support, DOD's Technology Reinvestment
Program, and Operations Requirement Development Manager within JSC
Exploration Program Office (Space Exploration Initiative proposed under George
H. W. Bush).
   Since retirement, he has joined BioDri Technologies / Space Legacy LLC as
technical advisor providing coordination and independent advice/recommendations
with regard to activities between BioDRI and NASA. In this capacity, Mr. Bell has
acquired a degree of understanding about antimicrobials and their potential space
applications and issues.

Chapter 14

Inflatable Habitats for Lunar Base Development

Eligar Sadeh, Ph.D.[1] and Willy Z. Sadeh Ph.D.[2]

Introduction

A human-tended base on the Moon serves as a stepping-stone for human
exploration and development of space beyond the International Space
Station. Achievement of this goal in the twenty-first century depends on
the development of low-cost and lightweight structures capable of
accommodating humans and supporting life on the lunar surface. One type
of structure suitable for the unique conditions present on the Moon is an
inflatable structure made of a thin carbon reinforced membrane. The
inflatable structure concept consists of nominally identical modules, which
incorporate a pressurized framing system to support structural geometry.

---

[1] Eligar Sadeh is President and CEO of Astroconsulting International LLC. His bio
   may be found at the conclusion of this chapter.
[2] The late Willy Z. Sadeh was the Founder and Director of the Center for Engineering
   Infrastructure and Sciences in Space and Professor of Space Engineering at Colorado
   State University. His bio may be found at the conclusion of this chapter.

The development of such inflatable structures suitable for use on the lunar surface is feasible, desirable, and economical.

The scenarios and concepts of inflatable habitats for lunar base development discussed in this chapter are a result of original and pioneering work conducted by the late Dr. Willy Z. Sadeh and Dr. Eligar Sadeh from 1995 to 2005. Past partners in this work include NASA Jet Propulsion Laboratory, NASA Johnson Space Center, Lockheed-Martin Corporation, Orbitec Technologies Corporation, Colorado Space Grant Consortium, and the former Center for Engineering Infrastructure and Sciences in Space at Colorado State University.

Since a "near" vacuum prevails on the Moon, habitats on the lunar surface are pressure vessels. Thus, the pressure of the internal artificial atmosphere within a habitat is its dominant structural loading. The main requirements for design of a lunar structure are selection of the proper materials and optimization for the specific loads. It must be easily deployable, durable over time, and entail minimum maintenance requirements. Modularity of the structure is required for lunar base expansion needs and for combining base functions, such as living quarters, laboratories, and manufacturing facilities, within a single habitat.

Inflatable modules provide for a straightforward design that results in an efficient and economical structure from both a transportation and deployment viewpoint. Lunar regolith shielding protects crews and contents. Figure 1 depicts the concept of an inflatable habitat.

This chapter highlights development aspects of inflatable habitats for lunar base development. The first part discusses scenarios for lunar base development and the role therein of inflatable habitats. Following this discussion, system requirements for inflatable lunar habitats are identified and defined. Third, geometric definitions and load conditions of inflatable habitats are briefly evaluated. Lastly, single to multiple inflatable module deployments and configurations are considered.

SCENARIO FOR LUNAR BASE DEVELOPMENT

The scenario for lunar base development, outlined below, is based on continuous expansion in mission requirements encompassing four development stages. These stages span from simple encampments to a human-tended mature base. The development stages include:

(1) Exploratory;

(2) Pioneering;

(3) Outpost; and

(4) Mature base.

Figure 1. Concept of Inflatable Habitats on the Lunar Surface.

The exploratory stage involves discovery about the lunar environment to gain greater knowledge about how humans can safely live and work there. This stage consists of unmanned robotic surveys and short-term human missions. Examples of such an effort involve robotic probes that NASA has sent to the Moon, the Apollo Moon landings, and the International Space Station as to its utility in understanding human adaptation to the space environment – physiological effects of microgravity, exposure to cosmic radiation, and sustainability of closed life support systems.

Objectives of the exploratory stage focus on the use of robotics for site selection for exploration, and the establishment of a future base encampment, sampling and testing of regolith and planetary subsurface materials, geochemical assessment, gathering of site topographic data, measurement of radiation and micrometeoroid impact, seismic and gravity data, testing of remote sensing and communication systems, and unmanned testing of lunar rovers.

The pioneering stage deals with building upon the knowledge gained in the exploratory stage to establish a human-tended base encampment. Two fundamental goals are essential to this stage: operational testing of human habitat deployment; and testing of in-situ resource utilization. The deployment and testing of infrastructure enabling technologies, like inflatable habitats, serve as a precursor for the eventual deployment of more permanent human habitats.

The objectives to realize more permanent human habitats include further expansion of the tasks outlined in the exploratory stage. This entails regolith handling and moving operations, robotic construction and mining, equipment testing, use of solar energy and batteries for energy storage, introduction of nuclear reactor for power generation, testing of heat rejection systems, initial tests of controlled life support system plant growth, intensive geological assessment, testing of robotic extraction of lunar resources, and initial manned exploration trips with lunar rovers.

The outpost stage marks the transition from an encampment to the establishment of a permanent infrastructure for achieving a self-sufficient base. Crews of ten to twenty astronauts with duty tours to the Moon of up to six months are envisioned here. The major task for this stage is the deployment and construction of an operational lunar biosphere. This entails life support outfitting and shielding of the human habitats deployed in the pioneering stage.

Life support outfitting is accomplished by integrating engineered closed controlled ecosystems technologies into the habitats, such as physical/chemical and air/water recycling systems, expansion of controlled life support system plant growth, and use of regolith as a plant growth medium. Radiation shielding and micrometeoroid protection is provided by a layer of lunar regolith.

The construction of an operational biosphere requires the development and use of a high-level of automation and robotics. Additional objectives to further develop the technology for a flexible, expandable, and permanent lunar base infrastructure include mining and extraction of in-situ resources as permanent ongoing activities, initiation of manufacturing techniques in lunar hypo-gravity, construction of photovoltaic solar farms for energy production, and extended manned rover exploration trips.

The mature development of a base is the most advanced stage before settlement. A mature base with a permanent population of up to fifty astronauts with duty tours of up to one year is the scenario at this stage. The primary goal of this stage is to establish a permanent lunar infrastructure capable of reaching an acceptable level of self-sufficiency. This implies that the infrastructure is capable of sustaining closed life support systems. Objectives of the base are to bring the tasks outlined for the outpost stage to the level of continuous activity that involves the manufacturing of products using lunar indigenous resources for export to Earth, and an expansion of the photovoltaic solar farm for energy production.

Other tasks at this stage in development include the construction of an astronomical observatory on the Moon and a lunar space port to enable robotic exploration of the outer solar system and beyond. An important aspect of this stage is to apply the knowledge and technologies developed in the base stage to enable human missions to Mars and to eventually establish a Martian base. Ultimately, the transition from a mature base stage to lunar/Martian settlements takes place when the base is economically viable and self-sustaining, i.e., independent and autonomous from Earth. The end point of this stage is the establishment of a "humanity extended home" in space.

The inflatable habitat concept plays a critical role to realize the scenario for lunar base development. It is a suitable technology to mature the pioneering stage that constitutes the United States Space Exploration policy and the return to the Moon efforts therein. Further, technical simplicity, economics, modularity, and flexibility for expansion establish inflatable habitats as a "technological-bridge" to extend from pioneering to more mature base concepts; ultimately, putting into place the infrastructure needed for settlement of the Moon.

SYSTEM REQUIREMENTS

System-level requirements for constructing, utilizing, and maintaining an inflatable facility on the surface of the Moon are identified and briefly discussed below.

MASS/VOLUME
Ensure useable mass and volume of the inflatable module to allow it to be transported economically to the lunar surface for deployment. The inflated volume needs to be adequate for the human crew to live and work for extended periods of time, and to accommodate all the necessary hardware, life support, and scientific equipment.

ECONOMICS
Produce an economically viable structure that can efficiently support the operations it was designed to accommodate— crew habitat, scientific laboratory, greenhouse, and storage facility.

CONSTRUCTION
Allow in-situ construction and assembly, and expansion of the inflatable structure into multi-module configurations.

STRUCTURE
Support internal and external applied loads through the structural components that are deployed.

ENVIRONMENT
Function within the lunar environment for extended periods of time.

SAFETY
Establish safety and reliability of the structure to support equipment and human occupants.

OPERATIONS
Facilitate maintenance of the structure once it is in place.

GEOMETRIC DEFINITION AND LOAD CONDITIONS

Structural components are designed from flexible carbon reinforced membrane materials. The framing system members are pressurized to provide stability of the structure whenever the module is deflated and to sustain the module shape. A thin web spans the arches and columns to provide a continuous load path for the wall, roof, and floor membranes, and to prevent geometric deformations. An elevation view of the module is presented in Figure 2 and in wire frame in Figure 3.

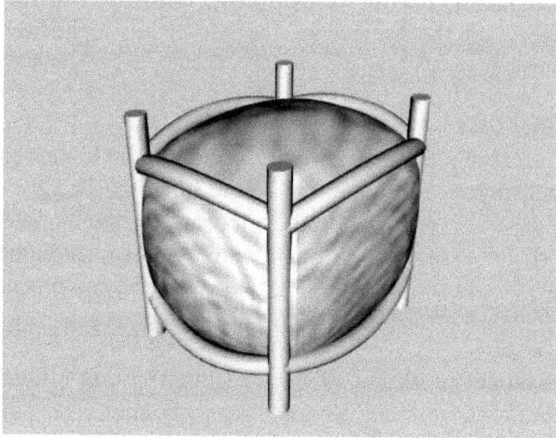

Figure 2.
Surface Elevation View of Inflatable Module.

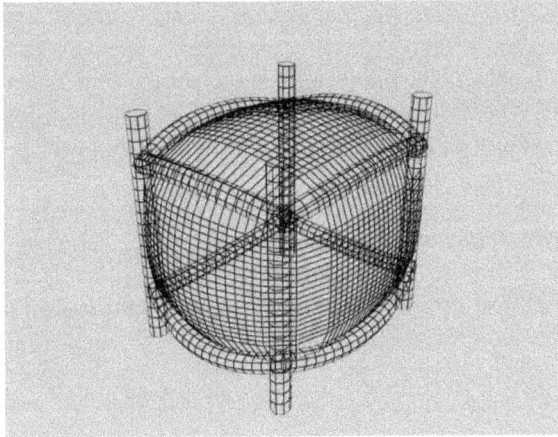

Figure 3.
Wire Frame Elevation View of Inflatable Module.

A candidate structure that was investigated by Willy Z. Sadeh can be comprised of a composite Kevlar flexible membrane. Minimum membrane thickness for structural integrity is 0.38mm for the roof, subfloor, and web, 0.64mm for the sidewall, and 0.76mm for the arches and columns. The total weight of a single module without any hardware or supporting equipment is less than 300kg. Overall module height is 6.55 meters, and the column spacing is 6.1 meters from center to center.

Loads on the structure can be divided into external and internal loads. External loads are largely a function of the environmental conditions on the Moon. Internal loads depend on the structure material, outfitting, and function. Weather loads are not a concern on the Moon. Therefore, the only external load to be considered is a regolith shield layer for protection against cosmic radiation, temperature extremes, and micrometeoroids. Internally, the primary load is the module internal pressure. The framing system pressurization is both a source of structural support and an imposed internal load.

The deployment sequence for an inflatable structure begins with pressurization of the framing system. This opens the structure so that internal outfitting may be installed. Once the internal fixtures are in place, airlocks are sealed and the module is pressurized. The regolith shield of minimum 3 meters thickness is then put in place by a robotic front-loader in the initial stages. In a more advanced stage, regolith can be transformed into bricks, i.e., the equivalent of lunar concrete, for shielding.

The adjacent membranes and the footpad reactions in response to gravity impose the first load condition. In a single module, the arches act as compression members forming ridges between the sidewall and the roof/subfloor as they are pulled inward towards the center of the module by the membranes. Similarly, the columns are pulled inward by the intersecting wall membranes. This force depends upon the curvature of the adjacent membranes, as a flatter membrane results in a larger force. The sidewalls assist the columns in holding the roof and subfloor membranes together by partially equilibrating the vertical loads imposed by the internal pressure.

A second load condition is based on the density and thermal characteristics of lunar regolith used for shielding. Based on the average lunar regolith density of 1750 kilograms per meter cubed, and accounting for the different gravitational force, this yields a regolith load of about nine kilopascals for three meters of regolith shielding, which is the minimum needed for effective purposes. This load is easily supported by an inflatable structure with internal pressures near sea-level here on Earth.

DEPLOYMENTS AND CONFIGURATIONS

Deployment and possible configurations for the inflatable habitat structure are visually shown in this section. Figures 4 and 5 visualize how a module inflates, how airlocks are incorporated into the structure, and how a cluster of four modules can be configured.

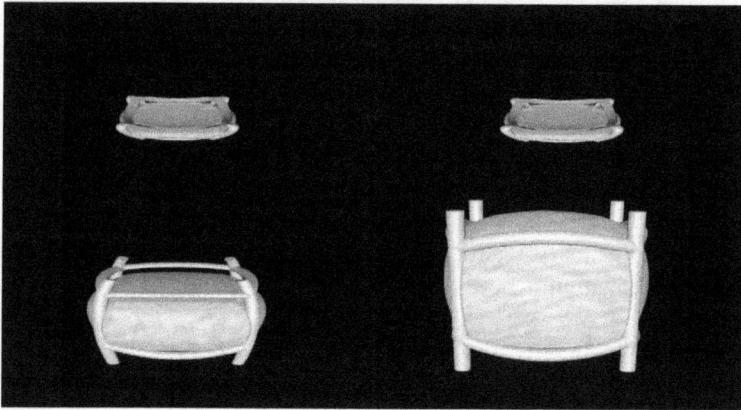

Figure 4.
Inflation of a Single Module.

CONCLUSIONS

Modular inflatable habitats for use3 on the surface of the Moon enable lunar base development in a number of ways. First, inflatable habitats serve as "critical enabling technology" for lunar base development. The basic system requirements for inflatable habitats allow for successful mission operations. Of particular note are the inherent economics, feasibility, and modularity of inflatable habitats.

Second, the concept is technically feasible. Prior work on inflatable habitats carried out by NASA and current work on the concept by Bigelow Aerospace can be leveraged to develop feasible concepts for the lunar surface. Bigelow Aerospace advanced the technology to a technology readiness level (TRL) of 7, which is where an actual system is complete and flight qualified. This mature state proves the technical feasibility of the inflatable concept.

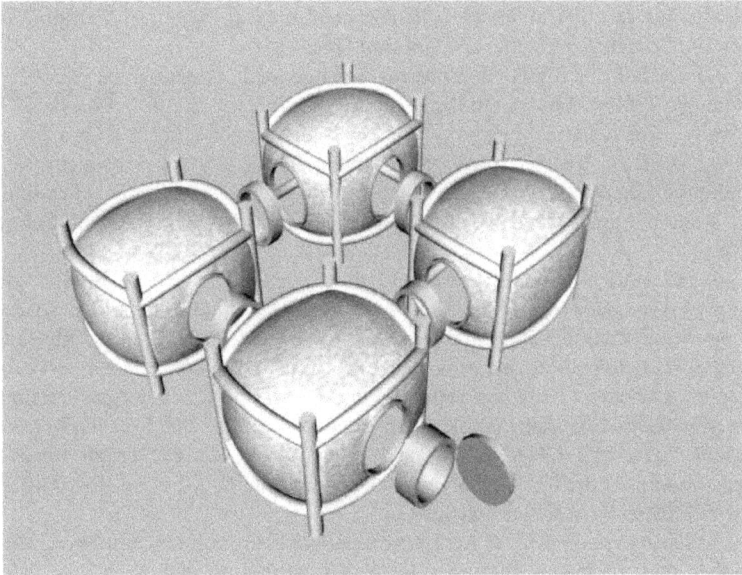

Figure 5.
Four-Module Configuration with Airlocks.

Lastly, the concept of inflatable habitats is politically feasible. If the politics do not fly, the hardware never will. The economics of the technology and the maturity in Technology Readiness Levels provide an attractive approach for lunar base habitats that can be realized within projected budgetary trends, and that can be developed and deployed on schedule.

•••

ELIGAR SADEH, PH.D.

Eligar Sadeh, Ph.D. is President and CEO of Astroconsulting International LLC, which provides consulting solutions, forum discussions, research studies, professional education, workshop development, editorial services, and language translations for the military, commercial, and civil space sectors in the United States and the global community. Clients and partners of Astroconsulting International include: Eisenhower Center for Space and Defense Studies at the United States Air Force Academy; Academy of Program, Project, and Engineering Leadership at NASA; American Institute of Aeronautics and Astronautics Professional Development; Center for Space Studies at the University of Colorado, Colorado Springs; Astropolitics of Taylor and Francis Routledge; and Imprimis Inc.

Sadeh holds a Ph.D. in Political Science with emphasis on Science and Technology Policy, Space Studies, and Environmental Studies, Colorado State University. He holds a Masters of Arts in International Studies with a focus on security, conflict, and cooperation, Hebrew University of Jerusalem, and a Bachelor of Science in Aerospace Engineering Sciences with specializations in space systems engineering and bioengineering from the University of Colorado, Boulder.

From 2001 to 2008, Sadeh held professorships in Space Studies at the College of Aerospace Sciences of the University of North Dakota, and in Space and Defense Studies at the United States Air Force Academy. Prior to this Sadeh, worked for Lockheed-Martin as a Space Systems Engineer. Sadeh serves as a Research Associate with the Space Policy Institute at George Washington University, Editor of Astropolitics published by Taylor and Francis Routledge, Editor of Space and Defense, journal of the Eisenhower Center for Space and Defense Studies at the US Air Force Academy, and on the Editorial Board of Space Policy published by Elsevier Science.

Sadeh has published in Acta Astronautica, Astropolitics, and Space Policy, and other peer reviewed journals. He is a contributing author to Space Power Theory Project sponsored by the National Defense University, a Space and Defense Policy textbook edited by the United States Air Force Academy, and he published a book in 2002 entitled Space Politics and Policy: An Evolutionary Perspective.

Sadeh holds adjunct professorships with the University of Colorado, Colorado Springs, Embry Riddle Aeronautical University, and the American Military University to instruct space-related academic courses. In addition, Sadeh instructs professional development courses in space policy and space management for the American Institute of Aeronautics and Astronautics, United Space Alliance Training Academy, and the Academy of Program, Project, and Engineering Leadership in the Office of the Chief Engineer at NASA.

Sadeh is a recognized expert in space policy and law, space studies, space systems management, and space technology development with sponsored research, and invited interviews, presentations, and testimony for, among others: National Science Foundation, NASA, Taylor and Francis Routledge, Eisenhower Center for Space and Defense Studies, National Defense University, Tauri Group,

Congressional Quarterly, US Government Accountability Office, World Affairs Council of America, History Channel International, and National Public Radio.

WILLY Z. SADEH, PH.D.
13 OCTOBER 1932 -12 MAY 1997

Dr. Willy Z. Sadeh was a dedicated and passionate scholar of space engineering and space sciences who believed that "any civilization that does not challenge the impossible is doomed to fail... and the impossible for our civilization is the human conquest of the infinite space frontier."

Sadeh was known internationally for his pioneering work in basic and applied space engineering and space sciences research with particular emphasis on the development of a human-tended lunar base. His research concentrated on inflatable structures for a lunar base, waste management systems in a lunar base, physical/chemical and bioregenerative life support systems for a lunar base, fluid management in a closed plant growth chamber in space, conversion of lunar regolith into a friendly soil, radiation protection in space, materials for space structures, and space policy on modeling international cooperation for space exploration.

As a professor of space engineering, Dr. Sadeh founded the Center for Engineering Infrastructure and Sciences in Space (CEISS) at Colorado State University in 1987 and served as center director the past ten years. He built the center into one of the premiere graduate level programs in space engineering and space sciences. The center promoted exploration and development of space, and coordinates interdisciplinary research, educational programs, and cooperative projects with academia, industry, and government in the emerging areas of space civil engineering, space life sciences, and space policy. Dr. Sadeh developed five novel academic options in Space Civil Engineering, Space Agricultural Sciences, Space Biomedical Sciences, Space Biology, and Space Policy at both undergraduate and graduate levels under the NASA Space Grant College Program.

He earned a B.S. degree in Aeronautical Engineering (1958) and an M.S. degree in Mechanical Engineering (1964), both from Technion-Israel Institute of Technology, Haifa, Israel; and a Ph.D. in Engineering with major in Fluid Mechanics (1968) from Brown University. He worked as a Forecaster for the Israel Meteorological Service; was a Structural Engineer with Générale Aéronautique Marcel Dassault, Paris, France; was a Research Engineer with the Aerothermic Laboratory, Rarefied Gas Dynamics Division, Centre Nationale de la Recherche Scientifique, Paris, France; was a Turbomachinery Research Engineer with Desalination Plants, Tel Aviv, Israel; and was a Senior Hydrodynamic Research Engineer in-charge of the Solar Pond Division, Negev Institute for Arid Zone Research, Beersheba, Israel.

He was on the teaching and research staff at the Technion as an instructor from 1962 to 1964. At Brown University, he had a University Fellowship in 1964-65 and was a Graduate Research and Teaching Assistant from 1965 to 1967. Dr. Sadeh joined Colorado State University in 1968 as an Assistant Professor, became an Associate Professor in 1970, and a full Professor in 1976. In 1975-76, Dr.

Sadeh was a Visiting Professor at Tel-Aviv University, Tel-Aviv, Israel, and a Visiting Senior Scientist at Cambridge University, Cambridge, United Kingdom.

Dr. Sadeh published 244 scientific and technical papers and reports, and was involved in over 300 professional conferences presenting papers, and chairing and organizing sessions. He received 39 awards in recognition of outstanding achievements and professionalism in education, public service, and research. The more notable awards include 1996 election as a fellow to the American Association for the Advancement of Sciences (AAAS) for pioneering and original research in space engineering and space life sciences, particularly for research in support of the development of a lunar base; 1996 American Astronautical Society (AAS) Victor A. Prather Award for contributions to the advancement of space engineering and space life sciences; 1993 Colorado State University Sigma Xi Honor Scientist; 1992 International Astronautical Federation (IAF) Frank J. Malina Astronautics Medal for demonstrated excellence by an educator in promoting the study of astronautics and related space sciences; 1990 and 1985 Colorado State University Engineering Dean's Council Faculty Award of Excellence in Engineering Science; 1978 election as an associate fellow to the American Institute of Aeronautics and Astronautics (AIAA); 1972 Engineering Science Vernon Dishman Memorial Top Professor Award; and 1971 NASA Certificate of Appreciation in recognition of his contributions to the NASA Technology Utilization Program.

Dr. Sadeh served on the IAF Education and Space Exploration Committees; and the International Academy of Astronautics (IAA) subcommittees on International Exploration of Mars and International Exploration of the Moon. He was on the Board of Directors and Executive Committee of AAS; was AAS Vice-President, Technical; served on the AAS Space Station, Space Exploration, Education, and International Programs Committees; and was chair of the 1996 AAS National Conference. He served on the AIAA Space Systems, Systems Engineering, Life Sciences and Systems, and Nuclear Thermal Propulsion Technical Committees, and the System Engineering Working Group; the AIAA Student Activities, National Region and Section Activities, Academic Affairs, and International Space Year Standing AIAA Committees; and was a member of the AIAA Rocky Mountain Section Council and Chairman of the council's Education Committee.

Sadeh was an international advocate for space exploration and development who helped to establish the engineering and scientific knowledge necessary for the development of a human-tended lunar base and human exploration of Mars. He was a true explorer and a husband and father who considered family as important as his work.

A Willy Z. Sadeh Graduate Student Award in Space Engineering and Space Sciences was established in cooperation with the AIAA Foundation to honor the life and accomplishments of Dr. Sadeh. The goals of the Sadeh award were to: (1) to support space studies research to enrich our understanding of the engineering, scientific, and sociopolitical aspects of space exploration and development of space; and (2) to expose graduate students, i.e., our future engineers, scientists and decision-makers, to the international astronautical community. The Willy Z. Sadeh Graduate Student Award was awarded annually to a Masters or Ph.D. student specializing in space studies at any accredited college or university in the world from 2000 to 2010.

Chapter 15

Sustainable Space Exploration and Space Development:

A Unified Strategic Vision

Feng Hsu, Ph.D.
Senior Fellow, Aerospace Technology Working Group
And
Kenneth J. Cox, Ph.D.
Founder, Aerospace Technology Working Group

The views expressed in this chapter represent those of the individual authors and of the Aerospace Technology Working Group, and are not necessarily the views of other organizations with which the authors may be affiliated.

A detailed version of this chapter is available to be viewed and downloaded as a white paper at www.atwg.org.

Introduction: The Need for a New Vision to Transform America into a Spacefaring Nation

This chapter recommends that a strategic, Unified Space Vision (USV) for comprehensive human space exploration and space development in the 21st century be established as the policy of the new U.S. Administration. The proposed Vision should replace the current Vision for Space Exploration

(VSE), which has been pursued via the NASA Constellation program since its announcement by George Bush in early 2004. We believe that the USV should serve the long-term economic, diplomatic and exploration interests of this nation and others around the globe.

There have been extensive discussions in the public and within the space-science, industry, and technology communities regarding the effectiveness of the current Vision for Space Exploration (VSE), and its proposed implementation as set out by the Bush administration. More than 5 years have passed since its announcement in early 2004, and it has become increasingly apparent that the rationale behind the formulation of the VSE and its implementation are problematic, and perhaps lack strategic merit.

In our view, the VSE suffers from these shortcomings:

1. Because there was not an open, well-informed debate involving the space and science communities, policymakers, and the general public, the current approach lacked the thorough review necessary to a strategic space policy statement. The proposed plan should have been scrutinized through hearings to engage the American public and politicians in the underlying thinking process.

2. The budget needed to fulfill the VSE far exceeds the resources currently available, as indicated by a recent GAO report. Consequently, budgets have been reduced in other important NASA programs such as Earth monitoring, space science, and robotic exploration. Compounding this problem, the program has not been fully funded.

3. From a strategic perspective, the VSE falls short of addressing the national and international needs and opportunities for space development. For example, the plan did not include a substantive process for international participation and support, which should be addressed in a comprehensive manner.

4. The current thrust of the plan to return humans to the Moon and to build a lunar outpost lacks political resonance. The American public and their representatives in Congress have shown little interest in supporting such a costly "Apollo-all-over-again" national program.

It is our view, then, that a new, unified vision is needed, one that fulfills the clear need to support both sustainable space exploration and the development of space commerce. What follows are the outlines that we suggest for such a Vision and its accompanying Policy implementation.

A Unified Vision for Concurrent Space Exploration and Space Commerce

Four decades ago NASA achieved inspiring successes in the Apollo era of Moon landings. Unfortunately, these successes have been followed by the frustrations of a number of compromised programs, cost overruns, and project cancellations.

NASA evolved from NACA, a civilian agency set up in 1915 to advance aviation technology, with a particular focus on serving R&D needs for the military. It was converted into a space agency in 1958 specifically for the purpose of winning the Space Race. Its governing paradigm emerged from its military roots, and it never developed the capability to address long-term strategic goals beyond Apollo, the Cold War, or the Space Race. Today's NASA is compromised by fierce turf battles among the ten NASA field centers, which results in damaging competition for the funding each needs to survive.

So while NASA accomplished exactly what it was designed to achieve with Apollo, its mixed performance since then may be a consequence of the same forces that made the agency so successful during Apollo. At its inception, the agency was not set up to envision, create, or manage a long-term development process, but this is exactly what is needed today. Our current strategic interests clearly lie in ensuring our national security by enhancing our leadership in science, exploration, and technology development, and in promoting world peace for sustainable human development.

We therefore suggest that America's space enterprise should be guided by a new entity chartered specifically to take on the concurrent challenges of fulfilling the strategic goals of both space exploration and space commerce.

Through millennia of history, humanity has achieved the "horizontal" exploration and economic expansion around our planet. Now it is time, at the dawn of the 21st century, for humanity to embark on the outward expansion into space – not only to explore other planets, but just as importantly, to enable a wave of space-based economic, commercial, and industrial development on the high frontier.

The history of human economic development shows that the development of new frontiers has frequently triggered significant economic growth. For example, the opening of commercial air transportation has advanced in all aspects, and has contributed tremendously to the world economy and modern civilization. This suggests possibilities for the next giant leap forward in humanity's economic and commercial expansion into low Earth orbit.

We therefore propose a strategic unified vision that addresses both space *exploration* and space *commerce*. It has four critical strategic components:

1. A program of sustained, affordable space exploration activities should be directed toward known and unknown planetary destinations beyond the Earth-Moon system.

2. An aggressive program of technology development should produce an affordable, low-Earth orbit (LEO) space-transportation infrastructure that would support sustained space science and exploration endeavors in and beyond LEO, and also enable rapid economic and commercial expansion into the Earth-Moon system.

3. The full participation of international partners, together with the private sector, would constitute a collaborative endeavor to address humanity's profound energy and climate change challenges, and would serve as a catalyst to foster world peace.

4. Given the likely significance of space commerce, it becomes evident that restructuring and realigning NASA's role is necessary to maximize its contribution to technology R&D and space exploration, and that a separate government entity may also be needed to effectively coordinate and promote the development of the economic infrastructure for LEO space and beyond.

Together, these elements constitute what we mean by a "strategic vision for unified space exploration and space commerce."

A CRITICAL PATH FOR ACHIEVING THE UNIFIED STRATEGIC VISION

Although an efficient and smoothly functioning NASA is critical to the success of the nation's space exploration programs, NASA and its efforts in manned and robotic planetary science should represent only part of the larger picture of America's activities in space. Even with adequate reform in its governance model, NASA may not be the right institution to lead or manage the nation's business in space development – because the programmatic principles and management culture appropriate for managing R&D projects in space exploration are fundamentally different from the principles and organizational culture that are needed to manage space commerce.

Human space commerce activities, such as development of affordable launch vehicles, RLVs, space resource development, space-based solar power, space tourism, communication satellites, and trans-Earth or trans-lunar space transportation infrastructure systems should be pursued according to a specific cost-benefit perspective subject to fundamental principles such as profitability, sustainability, and market development.

In contrast, space exploration, by its nature, involves human scientific research and development (R&D) activities suited to exploring the unknown, including "pushing the envelope" to reach new frontiers and taking higher risks. These activities enjoy government and public support, and are appropriately supported by taxpayer contributions.

NASA should therefore emulate the successful U.S. national research laboratories to focus on exploration, planetary research, scientific discovery, and technology development. In parallel, we propose that eventually a cabinet-level effort in space commerce should take charge of the government roles in supporting and incubating space-based industrial capabilities, including the development of a cost-effective transportation infrastructure.

The key role of the Space Commerce Agency (SCA) should be to encourage strong government-business partnerships, much like the current NASA COTS program, but at orders of magnitude of increased scale. The SCA should work with the existing nascent space industry and with established private sector firms to promote space infrastructure development. This will directly benefit the US and world economies by bringing returns to taxpayers, not just by creating more high-tech jobs, but also by supporting NASA in its pursuit of more ambitious space exploration programs.

The framework of the USV calls for key space exploration activities to be pursued through the following critical path for affordable and sustainable space endeavors:

1. As noted, an element key to all space efforts is the development of a comprehensive transportation infrastructure to serve the needs of both NASA's exploration agenda and the broader economic and commercial expansion into the Earth-Moon orbit systems. A cost-effective, heavy launch vehicle and affordable LEO transportation infrastructure is essential, and should be the task of highest short-term priority. It is crucial not only for supporting all strategic space exploration goals, but also for space-based commercial development. Crew and cargo transport and launch vehicle systems must address top-level requirements of low-cost, low system complexity, and must achieve aircraft-like reliability, maintainability, and operability.

2. We should then develop an international Fuel-Depot and Orbital Staging or Service point (station) in the LEO environment to support and service commercial space-transportation traffic, including space tourism, Lunar and Earth orbital transfers, and commercial satellite services.

3. We should also support the establishment of space port infrastructure in several locations within the U.S. and around the globe to meet the emerging demand for increased commercial launch and space-transport economic activities.

4. The international community and industry should be fully engaged in the effort to establish an international presence for lunar science and exploration. We have the opportunity for space development to be a strategy not only to strengthen relations with our allies, but also for enhancing mutual understanding and diffusing confrontation. We must avoid provoking a new Space Race, as it carries a high risk of getting everyone engaged in a lose-lose combative cycle.

5. Manned missions should utilize the Sun-Earth L2 libration point as a staging point for a mission to the Mars moon Phobos, followed eventually by manned missions to the surface of Mars. To achieve these goals, the U.S. should develop a Deep Space Habitat, which would be a module or station beyond low Earth orbit, complete with artificially produced gravity, to be used as a staging point at libration points such as the Moon-Earth L1 or Sun-Earth L2, or which would be placed in to orbit at various NEO destinations.

6. We also recommend an R&D effort and demonstration projects on space-based solar power (SBSP), which offers great potential for electric propulsion and power resources that can be utilized for deep space exploration missions. More importantly, its key technology components can be shared or used by many other space applications, including future supply of baseload power from space for terrestrial electrical energy demands.

7. It is also necessary to develop enabling space infrastructure observation and tracking capabilities for planetary defense. In particular, ground and orbital systems should be developed in close collaboration with international partners for monitoring, tracking and deflecting asteroids, comets, and other cosmic

objects in near-Earth orbits which threaten the safety of our home planet.

A NEW SPACE ECONOMY WITH TRANSFORMED GLOBAL COLLABORATIVE PARADIGM

History has brought humanity to the brink of an unprecedented era of crises, challenges, and opportunities for all of us. The current situations facing the world economy, energy resources, and global climate change certainly constitute dire threats, yet they also present enormous opportunities for humans to apply science, technology, and thoughtful economic development in the pursuit of meaningful solutions.

While some believe that humanity must solve our crises on Earth before we can expand into space in a successful and peaceful manner, we suggest the opposite. We believe that humanity is not going to solve all its problems here on Earth, and that we cannot create a utopia here, or anywhere, for this does not seem to be in our nature or in our destiny.

So we will venture into space not in pursuit of utopia, but rather for exactly the same reasons that our distant ancestors migrated from one valley to the next, and from one continent to the next: for adventure into the unknown, for resources, for freedom, and for better lives for ourselves and our descendants.

Recent events have shown that merely manipulating financial capital has failed to bring humanity a sustainable global economy, and so instead of fighting over the factors that limit and restrict human development on this planet, we must now expand our horizons, look upward and outward for resources, and embark on economic and commercial development of space.

Bold strategic visions supported by strong governments, and accompanied by leadership in technology and infrastructure development, have brought incalculable benefits to humanity. Examples include transcontinental railway systems, transcontinental highway infrastructures, and the creation of a global aviation industry, all of which stimulated progressive economic and cultural development in every nation. Similarly, the global energy and power infrastructures fueled the development of industrialization worldwide, and global communications infrastructures of wired, and then wireless, fiber optic, and satellite enabled voice and data are now essential to the functioning of human society and the global economy.

In addition to the economic and resource dimensions inherent in the space effort, there is also an important psychological dimension to consider. It seems to be a universal experience among astronauts that when they look back at our blue home planet from the depths of space, they feel what has been called the "overview effect," which is a profound natural bond for humanity accompanied by a desire to cherish one another. While humans, currently limited to a single-planet civilization, often feel threatened or compelled to fight for resources and living space on the surface of the Earth, we may anticipate that our psychological conditions may change as more people experience the overview effect as a result of expanding the human horizon outward into space.

CONCLUSION

To achieve a bold, strategic vision it is almost always necessary to think beyond the conventional paradigm. The indications are that the benefits to be obtained from space commerce could be enormous, and therefore we urge the Administration and Congress to support the development of a transformative space-based global economy.

If this were to happen, we envision that by the turn of the next century, year 2101, we could see vast economic activity spread across the near solar system, human habitats on many worlds, vast expansion of human knowledge, expertise, economic activity, and culture, and the transformation of the current economic crisis into the seeds of global opportunity that benefitted each and every human, and all of our descendants.

Economic development and the conquest of unknown geographic and technological frontiers have provided countless benefits throughout human history. What better strategic vision can there be for our future than the peaceful exploration and development of space for the benefit of all humans, alive and yet to be born!

•••

FENG HSU, PH.D.

Dr. Feng Hsu is a well respected world expert on risk and risk assessment. He is a former research fellow of Brookhaven National Laboratory in the fields of risk assessment, risk-based decision making, safety & reliability and mission assurances for nuclear power, space launch, energy infrastructure and other high integrity social and engineering systems.

He currently heads the NASA Goddard Space Flight Center risk management function, and is the GSFC lead on the NASA-MIT joint project for risk-informed decision-making support on key NASA programs, such as the GPM (Global Preciptation Measurement), LSS (Lunar Surface Systems) and the CxP (Constellation) etc. He was also a leading engineer/scientist in the Shuttle and Exploration Analysis Department at Johnson Space Center, SLEP and Shuttle upgrade trade studies etc.

Dr. Hsu serves on many agency and center expert panels supporting challenging SMA issues. He is co-chair of several international technical committees. He played key roles in the: (1) STS-107 (Space Shuttle Columbia) investigation team, (2) the RTF (Return to Flight) team, and (3) the ECO (Engine Cut-Off) expert team for the Discovery mission. Dr. Hsu has over 90 publications and is co-author of two books. He is frequently invited to be the keynote speaker in many international forums. His recent interests span from challenges to human space exploration to solar energy; and include global collective intelligence and risk based policy-making on emerging environmental and energy security issues.

Dr. Hsu holds a Bachelors degree in Applied Math, a Masters degree in Operations Research and Statistics and a Doctoral degree in Engineering Science. As a senior advisory member of the Aerospace Technology Working Group (ATWG) and a co-founder of Space Development Steering Committee, Dr. Hsu has been a strong advocate for Space Based Solar Power (SBSP) for years, and was instrumental in instigating the 2007 NSSO study on SBSP. Dr. Hsu has contributed whole-heartedly to the great human endeavor of harnessing solar energy for sustainable human development.

KENNETH J. COX, PH.D.

Dr. Kenneth J. Cox earned his bachelor's degree in 1953 and his master's degree in 1956 in electrical engineering from the University of Texas/Austin, and his PhD at Rice University in 1966.

In 1963 he joined NASA to develop the flight control system for the Little Joe II Booster Vehicle. Later, Dr. Cox became the Technical Manager for the Apollo Digital Control Systems, which included the Lunar Module, the Command Module and the Command/Service Module, the first spacecraft to fly with a digital flight control system.

From 1971 to 1987 he served as the Space Shuttle Technical Manager for the Integrated Guidance, Navigation and Control System, focused on developing new ways of utilizing a common set of digital computers to support all phases of flight.

From 1977 to1979 he supported the Apollo-Soyez joint mission between USSR and the USA in the Integrated Flight Control Systems.

From 1987 to 1995 Dr. Cox was Chief of the Avionics Systems Division, and directed early technology and digital system concept developments for the International Space Station. He was later was named the Chief of the Navigation, Control and Aeronautics Division.

From 1995 to 2003 Dr. Cox was Chief Technologist for the NASA Johnson Space Center, with the responsibility to promote government, industry and academia networking and collaborating for outer space related activities. He helped initiate an Avionics Technology Integration Group in the early '90s, the group that evolved into the ATWG. He also served on the Atlas Centaur Accident Investigation Board in 1997.

He retired from NASA in 2003 after thirty years of service to focus on ATWG and the peaceful development of space.

He frequently speaks at industry conferences. He conducted an AIAA Distinguished Lecture Series for six years, and other major conference speaking engagements include the World Future Society, the Creative Problem Solving Institute, Society for Automotive Engineering, the Science and Consciousness Conference, the Space Frontier Foundation, and many others.

He is recipient of the AIAA Mechanics and Control of Flight Award, AIAA Digital Avionics Award, and the NASA Medal for Exceptional Engineering. In 2007 Dr. Cox received prestigious Space Pioneer Award in Space Development from the National Space Society.

REFERENCES

1. Louis Friedman and Jacques Blamont. *A New Paradigm for a New Vision of Space*. The Planetary Society, Pasadena, CA. Nov. 2008.

2. William Claybaugh, Owen, K. Garriott, Michael Griffin et. al., *Extending Human Presence into the Solar System*. The Planetary Society, Pasadena, CA. July 2004.

3. George Abbey, Neal Lane, and John Muratore. *Maximizing NASA's Potential in Flight and on the Ground: Recommendations for the Next Administration*. James A. Baker III Institute for Public Policy, Rice University, January 20, 2009.

4. Feng Hsu and Ramny Duffy. "Managing Risks in the Space Frontier" in *Beyond Earth: The Future of Humans in Space*. Apogee Books, 2006.

5. Buzz Aldrin. *Fly Me to L1*. The New York Times, December 5, 2003.

6. Adriano Autino, et. al. *For a Politics of Support to Space,* The Space Renaissance Initiative, SRI, Nov. 2008.

7. Feng Hsu *"Harnessing the Sun - Embarking on Humanity's Next Giant Leap."* Proceedings of International Conference, Energy Challenges, Foundation For The Future, Seattle, March 2007.

ACKNOWLEDGEMENTS

The authors are deeply indebted to many of our colleagues at ATWG for their support and help in editing this paper. Special thanks to Mr. Rick Eckelkamp, Dr. Sherry Bell and Mr. Langdon Morris for their comments and suggestions in the preparation of this document. Our heartfelt thanks also go to Mr. Adriano Autino, Annie Bynum, and Dr. Raymond D. Wright of the Space Renaissance Initiative, and Amara D. Angelica.

FICTION

Chapter 16

There Be Humans There

Jamey Brzezinski, M.F.A.

Gianna Pittman entered the airlock, sealed the inner door, and started the pump. When the airlock had equalized to zero atmospheric pressure she checked her suit gauges, which showed no pressure loss over the mandatory 60 seconds. She checked the back-ups and, seeing an all clear, un-dogged the outer hatch. Moving out into the vacuum, she made her way to the small crater that served as her studio.

The crucible was intact and reaching optimum temperature in the bright lunar sunshine. The mixture of silicates, potash, and oxides was beginning to vitrify. Today's globule would be pink, and so she texted the notations she'd made to herself so she could later add the stats to her documentation.

Axel Rodriguez had been her mentor during her graduate studies back on Earth at the San Francisco Art Institute, where she'd studied from '83-'86. He had always emphasized the importance of documentation. As a pioneering artist in the genre, his initial projects 25 years earlier had inaugurated the first serious critical dialogues on space art. He had, under the aegis of NAASA and the NEA, created his first piece in 2058. He had also created quite an enormous controversy.

He had released over a ton of small, reflective, metalized Mylar chips in a narrow stream over a period of 90 minutes: one complete orbit. The small, but highly reflective chips had been intended to form a temporary ring around the Earth. Earthring, as the piece was titled, traced the trajectory of the launch craft, a third generation Russian shuttle under contract to the North American Aeronautical and Space Administration, in its orbit around the planet. Earthring was expected to only last a few days, as inertial and gravitational forces would slowly dissipate the chips—some falling to Earth, some being jostled outward and away. Totally unexpectedly, however, the ring and its residue lasted for more than six years.

At first, the prismatic effect of several billion tiny mirrors reflecting the sunlight as they tumbled slowly through the night sky, often reaching in a thin line from horizon to horizon, was very popular. The critics raved, calling it "the longest line ever drawn" and a "spectacular work in a totally new form." Sociologists noted that even the birth rate during that period was affected, as millions of young lovers reclined out under the summer sky, inspired by its beauty.

The first criticism came from the astronomical community. They had objected to the piece from the outset, citing the obstruction of an entire slice of space that would be obscured by reflected light. After Rodriguez conferred with them, promising to optimize the timing to create as little disruption in their studies as possible, they eventually decided to be "good sports," and to tolerate the interference in their observations and cataloging.

The environmental community was concerned about everything: from what they termed "space graffiti" scarring the beauty of the heavens; to the effect of the additional light on the navigational abilities of nocturnal moths.

When, after the second week, it became obvious that Rodriguez' assumption about how long his piece would endure was off by a factor of 300, both groups went ballistic in their condemnation of the project.

It turned out that the Mylar chips also disrupted certain wavelengths of radio frequencies, rendering a small, but nevertheless useful portion of the radio spectrum all but useless for communication or signal transmission. This, in turn, engendered suspicion on the part of some members of the international space community, especially the Equatorial East-African Union, whose program to establish its own lunar base was stymied -- as the affected frequencies were mainly those among the bandwidth assigned them by the United Nations Treaty for the Peaceful Use of Near-Earth Space, and its' Codicils of Cooperation. Some believed that "Rodriguez' Ring," as the popular press dubbed it, was an attempt by the first-world nations to limit opportunities for less wealthy, former colonies—even though those colonies had won their freedom from the European Empires almost a century before. Resentment dies hard in a post-colonial world.

Gianna finished texting her notes concerning the level of iridescence coming from the view-port on the crucible. The glass was now fully molten. She had to inspect the expansion hydraulics, which when released would launch her latest piece, Ovum #117, out from the moon. The fifteen minute inspection was grueling, requiring the removal of several couplings to access the viewing ports. This entailed applying her heavy inspection mandibles in awkward positions while taking care to avoid abrading her enviro-suit on the sharp-edged fittings, or on the rough lunar surface. Once inserted, the tool had to be rotated to ensure there would be smooth articulation of the release mechanism at launch.

When she was finished, she'd broken a sweat and had depleted 83% of her breathing stores. The Protocols of the Exterior, the rules that governed the activities of private citizens on the lunar surface, and granted her an environment certification, were absolutely firm. No less than a ten percent supply of breathable gas must be maintained at all times. The penalties for even one violation could include suspension or complete denial of Exterior Guild privileges.

She quickly stowed her inspection mandibles, figuring she would hurry to the exterior hatch with a three or four percent margin. Close, but not too bad—unless her logbook revealed a pattern of "frequent and questionable" close calls. For artists such as her, these restrictions were tighter than for the maintenance crews and engineering teams, who on average spent between five to seven hours on the lunar surface each day. Gianna's routine rarely exceeded an hour per day, but her grant required a launch each day that her apparatus was running, timed, of course, to the lunar cycle. It didn't provide any salary or stipend for an assistant, which left only her to inspect and operate the solar crucible in which her art works were prepared.

The principle was simple. The colony was located near the lunar equator. Gianna had developed the technology to focus the unfiltered sunlight available on the lunar surface with sufficient energy to make glass. As the materials vitrified into a globule in the enclosed crucible, a small amount of inert gas was injected into the super-heated material. Beginning with old prototypes that had been developed by Pyrex more than a century before, she had developed a glass capable of going from a molten state to a solid without fracturing; while withstanding an almost instantaneous temperature drop of over 3500 degrees. Her secret had been to infuse small bits of ferrous compounds into the glass. This enabled her, once the gas had been injected, to spin the globule magnetically at an incredible velocity in the almost frictionless environment.

Exposed to the freezing vacuum through the hydraulic operation, the spinning glass was annealed into an extremely hard and almost shatter-proof substance. In the instant before it solidified, the vacuum expanded the gas, creating a bubble whose diameter perfectly matched the crucible walls in a very snug fit, sufficient, in essence, to cork the bottle. It had taken Gianna hundreds of hours of calculation, trial and error, and experiments with the materials to determine just the right combination of factors to produce so precise a bubble in the extreme environment of the lunar surface. And it was an effect that could only be obtained on the lunar surface.

The pull of gravity, reduced to one sixth of Earth's, allowed the globule to spin freely during its expansion, which nearly eliminated the sag that has been both boon and bane to glassmakers for millennia. This allowed for a more perfectly symmetric sphere. The vacuum of the lunar exterior created the mechanics of expansion without the necessity of building a pressurized containment vessel. This, with the added plus of no atmosphere to transmit and contain heat, near absolute zero in the shadows, and the intense heat almost hot enough to melt lead in the direct sunshine, created an almost perfect environment for the creation of her Ova.

While this could also have been accomplished in the freefall of orbit, Max Barnett's failure to contain his molten bronze in the manufacturing bay aboard the International Space Platform in the early sixties and the near disaster it had caused; which cost several million dollars to repair, rendered all subsequent art projects nonstarters.

In applying for her grant, Gianna had hit on using the crater as her studio. The solid lunar surface beneath her crucible was stable, and if something did go wrong, as it had that once, the blast would be deflected upwards and outwards, contained by the crater's wall. While zero gravity would have been perfect, the lunar surface was as close as she was going to get.

And of course the lunar outpost offered the distinct advantage of abundant raw material. With plentiful silicates in the lunar soil, she only required small amounts of potash, soda, and metallic oxides to complete her formula.

The super-heated gas below the bubble built up enormous pressure, but it was held in check by the now hardened bubble. She then injected a few precious drops of water into the lower area, which flashed into steam at the same moment the launch hydraulics released. Like a cannon shot straight up from the lunar surface, out from its equator, the bubbles of colored glass were able to overcome the weak pull of the lunar gravity, as well as the distant Earth's pull, and proceed straight away on their journeys.

Since she always launched the bubbles while the moon faced away from the sun; their journey took them out towards the vast outward expanses of the solar system. It was the timing, synchronized with the lunar cycle, that led Gianna to call the bubbles Ovum #1, Ovum #2, etc. She titled the entire conceptual piece Ovulation.

Back inside the airlock, all was ready for the launch of #117 as she wiped the last of the suit talc and sweat from her skin and slipped on her interior coveralls. Hurrying down the tunnel to her control studio beneath the Earthward side of the crater wall, she quickly assessed the criteria registers. All indicators were a go.

This was her third functioning solar crucible. Each had been larger than the previous. Ova 1-27 had been a mere eighteen inches in diameter, 28-80 were over three feet, and now she was ready for the 37th ovum from the seventy inch pot. She'd resolved the issues with the crucible's ceramic make-up in the one meter pot. Number 27 had cracked the pot. Fortunately, explosions are silent in a vacuum, and it wasn't until several days of cleanup had passed that an inspector stopped by. Equally fortunate was that the inspector was Jeth Lambert, who was about three years older than Gianna, and who obviously had a crush on her.

She liked him too, and in their conversation he failed to notice the little telltale evidence that remained. Or perhaps, knowing how important this work was to Gianna, he overlooked it. Some bits from this explosion were blown as far as 35 miles away, and it was a minor miracle that no one and no thing had been damaged.

"Launch" or "Launching," as it was sometimes called, developed its own aesthetics for the artist and the viewer. For Gianna, it was the beautiful balance between process, materials, and metaphor that intrigued her. Although not a spectacular launch in terms of splash and fire, the smooth trajectories of the seventy-inchers were quite visible for weeks, so that at any one time up to a dozen Ova could be seen among the stars.

Of the 100 launched during the first year, all had survived, except number 27, which Gianna laughingly referred to as her 'miscarriage.'

The first generation of "Launched Art" had been larger, more spectacular, but somewhat bombastic. Chemical explosions, really no more than enormous fireworks, had been ignited at various orbit altitudes. These would expand outwards for tens of miles, and were visible from much larger portions of Earth that happened to be in shadow. "Detonating" was perhaps more accurate a term than "Launching." However, since the point of art is to be viewed, and each small fragment was directed outward, away from the planet, "Launched" or "Launching" seemed a more appropriate descriptive term than "Detonating."

These were extremely short in duration. The Italians and the Chinese, perhaps because of their long traditions in pyrotechnics, mastered these events. Ling Lin's Immediate Diaspora, which referenced the centuries of movement from China outward and especially the sudden migrations of the 2020's, was described as "a yellow, spherical tide of Asian thought, enhanced by power, entrapped by prestige in the wake of Chinese dominance in manufacturing and economics," in an article published in Artquorum, the international journal of record for new genres in the visual arts.

Similarly, Gianlorenzo Putinni's Fast Retreat, which expanded outward and then reversed, retreating back to its point of origin, was a metaphor for twenty-five centuries of Italo/Roman military, cultural, and imperial aspirations, which "became obsolete, like a withered iris, in the post European Union era." Although the "withered iris" simile was perhaps too florid for some peoples' taste, art critic Simon Selz penned the phrase with full awareness of what would soon become a world-wide rhizome blight that would go on to nearly wipe out several prominent iris species and devastate the urban landscapes of more arid climates. The species was about the only flowering ornamental that had been able to withstand the globally warmed environment in areas like California, Italy, and Spain, where they were very popular. A critic and amateur gardener, he also noted that its' curious, cruciform asymmetry did, in fact, look like an iris.

As the 21st century advanced, and payload costs decreased, more artists were "Launching." But the short duration of these works limited their scope and substance. This explained why Rodriguez' project was greeted so enthusiastically—at least at first.

Gianna's final touch, perhaps as a human being, a woman, and an artist, was also her true final stroke of brilliance: she always put a little of herself into her work. At first it was accidental, as a few strands of hair fell into the mixture and were too difficult to extract. Later, she consciously began to include bits of fingernail, tear drops (when her father had died on Earth around #14), and then drops of menstrual blood. Although she had been calling the pieces Ovum since the project's conception, the inclusion of her own genetic material added a deeper layer of meaning, both literally and allegorically.

Knowing that any organic material would be completely carbonized by the intense heat involved in the process did not dissuade her, she enjoyed the ritualistic aspects of what she thought would be a meaningless bit of matter: of no consequence to the work itself. She felt it also imbued the piece with a mysterious quality, a feeling that perhaps stemmed from her grandmother's mystical version of Catholicism.

As she calibrated the tracking cameras that would follow #117 for its first 24 hours, the time it took in the freezing vacuum of space for the piece to completely cure, and thus reassure her it would survive, she mused on the notion of the idea incarnate; how creation for the sheer joy of it was the work of only gods and humans, and how the artist in her could not be stopped any more than the inexorable and miraculous development of an embryo in the womb.

She was enormously satisfied that her work was pioneering a new genre, and in the process making allusions to the fundamental desire in all humans to create, all in a way that was distinctly female. As the womb is the crucible of life, her work was an allegory full of the feminine. The fire of passion resulted in the fusion of material elements into something with a "life" of its own; the ultimate launching of that creation into the universe

beyond our cozy home among the inner planets, and the subsequent sense
of the loss of that creation, gone from home forever—these were things that
all mothers had to endure.

She reflected on this as she stowed her tools and prepared for the next
day's work. As with mothers and their ova that become children going out
and on their away, her Ovum, her piece Ovulation and all its ramifications
would continue on and on, far away from their starting points on the
surface of the moon, with a destiny of their own. Like bottles cast upon an
endless sea, so too were her simple glass spheres, created with a message
inside, that touch of carbonized genetic material. Carbonized yes, but
redeemable as well, when the proper technology was applied.

We raise our children and then send them out into the world. Every
genius, every visionary, and every tyrant, murderer, and dictator started this
way as well. Sometimes the innocent and pure are seduced by evil in the
world and the results—sadly evident in 5000 years of human history—are
often catastrophic.

On the very day that Gianna drew her last mortal breath, 43
productive years later: after a life of accolades and success; of art projects
surpassing Ovulation in their power and brilliance; of family and friends
and lovers and detractors; of a life as rich and fulfilling as any human could
ever expect or even dream of, there were consequences.

On that very day, that very work of art, Ovum #117 was encountered
in the far reaches of space. And although Gianna's intentions were noble
and true to the best instincts of the human animal, she did not know that the
humans of Earth were only but one of many human species that had been
dispersed across the galaxy two hundred thousand years earlier. Arriving
on our pleasant blue planet as colonists, disease, tragedy, and the passage
of time had broken the link with our home planet and our cousins in other
star systems, marooning us in a backwater spiral arm of the galaxy.

She also did not know that other, non-human species considered her
own as pestilence, as vermin whose creative impulse was understood to be
disruptive to the good order of the cosmos. For millennia, they had moved
from planet to planet, tracking and hunting down human civilizations.
Already they had cleansed many planets of this creative infestation,
returning them to pristine status.

As they happened to pass by, out beyond the orbit of the giant gas
planets, they did not suspect that this system was also suffering from
human infestation.

Sightless, their communications through the void were based on
means that left them oblivious to the radio and light waves that Gianna, and
the rest of us, find so sublime. Their species had long ago developed
technologies that were far more advanced than our own, and were now so
ancient that the alien-ness of their existence precluded any understanding

of humanity's creative impulse because it precluded communication itself. But they were intrigued by the 70 inch glass sphere they encountered.

So curious were they that they brought their technology to bear in analyzing it, and they were able to retrieve and revitalize that small bit of chemical residue that had been Gianna's DNA. And this led them to the undeniable conclusion that caused them to change the course of their great vessel to retrace the trajectory of Ovum #117. They had encountered this problematic DNA elsewhere in their travels. It could only mean one thing.

Now they were filled with the excitement of the hunt. If put into what we think of as words, to them it could only mean one thing:

There be humans there....

•••

JAMEY BRZEZINSKI, M.F.A.

Artist Jamey Brzezinski, M.F.A., has shown his paintings, drawings, and prints in over 150 solo and group exhibits in the USA and abroad. His work is included in several significant public, private, and corporate collections. Brzezinski has been a Professor of Art at Northern California colleges and universities for thirty years, for the past sixteen at Merced College, where he was Chair of the Arts Division from 1996-2008. A former editorial cartoonist for an Oakland Weekly and Northern California Editor of Artweek Magazine he has also published over thirty technical, critical, and catalogue articles. He also plays jazz bass and divides his time between Merced, Mariposa, and Pacifica, CA.

"Art has always propelled technology and imagination on a species-wide level. Every manufacturing process evolved from the manipulation of materials by artists. Casting, stamping, and all manner of working metal, metallurgy itself from ceramic glazes, glass blowing, drills, chisels, stone masonry, chemical dyes all developed from the artist's quest for beauty millennia before they were used in the industrial sense now associated with them. The newest technology I teach is lithography, invented in 1793. Micro-processor technology relies on lithography to print the circuitry onto the surface of silicon wafers. The oldest technology I teach is that mid-wife of literacy: drawing. The world's oldest archaeologically provable profession, artists have been around between six and ten times longer than farmers. The history is fascinating, but then there's the future."

CHAPTER 17

THE KINGDOM OF ADZ

RICHARD E. ECKELKAMP

2105 A.D.
Neu Hoffen

Theresa, where is David?

He horsed out twenty minutes ago, Dad.

Good. I was beginning to worry that he had forgotten to service shutdown the subterranean reaper-mats. The night's cold could frost-crunch those silconites badly.

———

Watch it, Jeeves! How come you horses can see the difference 'tween dark lava of road and dark lava not of road? Sure hope that Claire 'membered to tell Dad that I was leaving. Having four sisters and me being the only son brings lots to do. Thought this was an equal society. Why are there still chore-bunches reserved exclusively for the less fairer sex? But, I am not complaining, Lord - just sociologically musing on the intricacies of our rural society, this rural planet, this incredibly special-to-me und- alles, place to live, platz zu leben.

Jeeves, I'm glad you're carrying me 'stead of vice 'vers. It's steep up this ridge. But what a view! Our family's 17000 hectares includes the region's second highest elevation. Only Mt. Carmel with its Kirche is higher. My grandfather, back on Earth, always told us of his dream of living in a beautiful valley like in Deutschland where his great-great grandfather lived, a valley with a small village surrounded by farms and woods and open spaces and peace and work and loving families, nestled mid the mountains, snow capped all – an idyllic existence, one that we experience here now, so far from Terra.

From up here the houses and roads all seem so tiny, a little bit above bare perception. Whoa, Jeeves. Shhh. Listen to the Kirche chimes. They sound much like the bell my grandfather six generations back purchased for his church on another Mt. Carmel in a then new land across from Europe long ago. The peels are inviting, making it easy to remember our true Father. I really relish this experience! How marvelous are Your works, O Lord. How beautiful to behold.

What a twilight! The stars coming out like diamonds in such a rich dark blue sky. Thinner atmospheres do indeed accentuate the lower spectra, shades of Raleigh scattering, here un-subdued by tropospheric aerosol pollutants. Our air is clean and crisp and cool and produces vivid sunset colors, hues of scarlet and orange and red in the cirrus layers giving way to violets and deep rich blues as our planet is put to its nightly rest. Peace is definitely settling in: Jeeves' hooves crunching on the darkened road, the feel of the damp rolling in, now the quieted calls of the evening birds, the appearance of the first of our three moons.

Oh Lord, I really love our life! What a great time to be alive. What a great planet to live on. We are rich in blessings: that Star Settler transport, nine years ago, bringing our Dad and Mom, Paulus and Mary Steyermeier, and us little youngers here to this new planet, Neu Hoffen, in the stellar outers, fifth from the star Darion, twenty three parsecs from our original Earth; our family growing now- not just Maria, Dymphna, Theresa, Claire, and me – now a new one on the way. Mom says it's a boy. Shoot, I'll be twenty-three before he gets here. I won't get much help soon. More blessings, Lord: our father being appointed to Prince Leonard's advisory council, the Roundtable; the fertility of this land for crops both above and below ground; the specialness of our society; the closeness of its families; the happiness of being able to work in a free and great land. Wow!

This blessing list seems endless. Our medical advances enable us to prevent physical problems and live much longer by making proper lifestyle decisions. Our grandfather, having recently immigrated from Earth, is one hundred thirty-seven. I think that he can outwork my Dad. Women can bear children into their nineties. The Human Genome Project outfalls give us methods of curing and even preventing most diseases. We're advancing well in all of the sciences - case in point: our Star Settler brought our family here in nine months - a contribution from the higher-ordered v/c terms ignored by the relativators in their quest for quadratic perfection. No sub-light for us.

Not that we don't have challenges. The Kingdom of Adz, ruled by Prince Leonard de Limousine, finds vigilance necessary for survival. The planetary elements are tough themselves, for example, we require extra efforts against the seasonal and nightly cold; but it's the human challenges emanating from the some of Neu Hoffen's other provinces that give us the most cause for concern. Some other parts of our planet are not as prone to peacefulness as are we. This has been both quite surprising and disappointing to us. Everyone who mounts a Star Settler craft takes an oath to abide in peace, wherever the destination. Why do some people ignore what they promise, break their oaths, and choose to live such lives? This kind of behavior is foreign to us. Our entire society here in Adz and on many of the surrounding provinces is deeply Christian, tolerant of others, hardworking, and friendly. Why would some wish to do us harm? God probably asks this same question every day.

Right now Dad's on his way to a Roundtable meeting at Mt. Carmel to examine what can be done to upgrade our defenses and our "military responsorial tools" in our pay-as-you-go society. We have electromagnetic sensor nets employing copious frequencies atop our higher places including this, our family's ridge. Down below we have more tools of detection. 'Tis quite costly for us to maintain both these and our weaponry – 'tis the devil's tax, it is. And we pay in other ways. Most of our energy production equipment must be underground to protect our ability to live, to farm, and to fight. Each of us takes our turn at watch, even my sisters of age. We have this valley, the highlands, and the surrounding plains to protect. Every adult is a part of the protection force – reminds me of Switzerland in yesteryear's Europe.

One blessing that we have in our Adzenian society is that these protective necessities do not dominate our attention, our psychic, or our activities. Protection service is just something each of us gives as a part of everyday life.

What does excite us, what does motivate us, what does lure us are our desires to expand our minds, develop the individual and collective gifts bestowed upon each of us by our loving Creator, better our society, and help each person and group that we encounter. We do all these things with passion. Our reasons go deep into ourselves, usually becoming private at a point where love, duty, and Love intersect.

We are a kingdom of the mind. To learn is a great joy for us. We all study, all learn, all partake of formal schooling from waking till laster-year age. We speak many languages: English, Chinese, Spanish, German, Astroen, Japanese, French, Jovian, Russian, Latin, Swahili, and others. We delight in discussions on physics, astronomy, genetics, sociology, classical arts, and other self-enhancing endeavors. We stretch our minds with games, puzzles, and reading.

We are also a kingdom of the body. Every individual engages in a complete complement of physical activities and exercises. Through these endeavors we continue to improve and develop our bodies till called home by our Creator.

My Father, like many others in Adz, requires each member of our family to wear gravity belts, causing our muscles to move and work as if they were still on Earth rather than enjoying the eighty-seven percent g-field that Neu Hoffen offers. He tells us that someday we may visit our family's home planet and we wouldn't want to be unable to enjoy all its fabulous pleasures because our bodies are weak. Those g-belts make it hard to walk, to work, sometimes even to breathe; yet, we all think that they are beneficial. It also happens that many Adzenian maidens take me bene notice - not a bad deal as my mate search continues. Muscles impress!

In addition, we are a kingdom seeking the spiritual. To love and to serve are our passions. To increase in self-discipline and virtue are delights for each of us.

We are a kingdom of interlocking families. Most of our social activities are centered around our families and das Kirche. We get together for conversation, food, games, parties, dances, cultural outings, nature adventures, and literary events. Nobody has any secrets. How could they? My father is one of eleven. All of his brothers and sisters with all their families are here in Adz. I and my sisters have thirty-five first cousins. There's a might more of seconds and thirds. I'm just thankful that there are copious entities of the female type who do not blood relate to me. My goal is to find that special one by the time that I have dated one hundred and eighty women (call me Phoenician). Finding the right one will greatly reduce the electron count in my schedule book, not to mention I might get more sleep.

Hey, Jeeves, we're clopping up the last leg of the ridge. I can see the reaps. They look like Stegosauria against the back-rock. Madre mia, listen to those brats moan. They creak like floundered windmills. I'm always amazed that those machines even work. They look like a cross between ...

The Continuation...

•••

RICK ECKELKAMP

Richard (Rick) E. Eckelkamp was raised on a dairy-wheat farm in a North Texas German farm community. His college education includes attendance at Assumption Seminary and St. Mary's University in San Antonio. He received a B.A. in physics from the University of Dallas with minors in mathematics, philosophy, theology, and foreign language, and has done graduate work in physics at the University of Houston. Before coming to Houston to begin his work at NASA, he worked as a physics and general science instructor at the University of Dallas, was an oil field worker, and performed atmospheric physics research at the Graduate Center of the Southwest (now UT at Plano).

At NASA from 1967-1987 in the Mission and Planning Analysis Division, he developed extensive experience in Earth orbit, lunar, and interplanetary guidance, navigation and control, engineering analysis, and flight techniques. He performed flight software verification and flight data production and performed Apollo through Shuttle flight control, serving as Shuttle console leads for both onboard and ground navigation. Rick was a major designer of Shuttle onboard navigation and fault management software, and developed ground control center software and techniques. From 1987-1995, he performed Mars and advanced lunar program software/hardware design, performed space station automated operations systems design, analysis, and prototyping, and served as the Systems Development Manager for the Space Station Freedom command and control system.

From 1995 to 2006 Rick worked in advanced life support command and control, served as the Software Integration Manger for the International Space Station Robotics, and performed systems engineering for exploration era robotics and integrated operations control. Since 2003 he has helped shape policy of the Aerospace Technology Working Group (ATWG) and serves as the head of its robotics forum. He was a principle participant in NASA's exploration content planning for 2004-2005. In October 2006, he retired from NASA to become an independent space consultant. In June of 2008, he became the Chief Executive Officer of the ATWG. Along with its founder and chief advisor, Kenneth J. Cox, he is transforming the ATWG into an organization that does both volunteer work and per fee tasks for industry and government.

Rick is an enthusiastic speaker on space subjects and inspirational topics. He has authored articles on faith and specific space technologies in magazines and scientific journals, and has written hundreds of NASA reports on guidance, navigation, and control; spacecraft command and control; and ground-based and vehicle-based operations. Rick has recently authored chapters in both Beyond Earth, The Future of Humans in Space, and Living In Space.

He is an Eagle Scout, a public speaker, a teacher of self-improvement and technical subjects, an independent businessman, a father of seven, a scoutmaster, and a religious educator. His hobbies include gardening, custom woodworking, camping, conversation, and travel.

CHAPTER 18

THE JOURNEY

MICHAEL F. HANNON

Sssssshhhhh......Boom!

Rattling, moaning, screaming metal resonates to the deadly winds and rippling ice as he wrestles with unfamiliar controls, led by near instinctive calculation somehow recalled from memories he never knew he had.

A sharp roar of ice particles slamming against the hull of his enclosed craft at times remind him of a war he never fought, or couldn't remember, yet the thoughts arise, toss, and wander too quickly to even seek recollection. Regardless, he can't dwell on such things while being driven by gusting winds of 100 or 200 mph which he could not possibly survive without the shelter of the shining alloy craft within which he has found himself in this moment of near shock, trembling from the sheer velocity of everything, including himself, as he flies over the vast planet of ice towards a gigantic sun dead ahead. The rattle and shudder of the wheel he holds desperately, both hands covered by strained tight gloves that belie white knuckles beneath, offering barely adequate protection needed in such a cruel yet beautiful world before him.

He does not remember who he is, or where he came from, but not that he has wondered at this point for more than a millisecond. He is bound to the instant by its urgency, as everything descends upon him at once, and for a moment he is consumed by an array of simultaneous emotions – fear, wonder, uncertainty, panic, confidence.... Then they swiftly trail away as though left behind like the scrapings of the edges of hard steel runner blades that grip the ice and toss aside the shards cut by them into a wind that takes them far away. He is alone, and seeing this, himself in the midst of this – all of it – brings a welling in his eyes that he struggles to dismiss. He knows he cannot succumb to such considerations now – he must wait – at what could easily be 100 mph as his craft shoots down the ice to what may or may not be somewhere, or nowhere. He can only wait.

After hours, he studies the hands on his wrist for a second, hoping that it doesn't stop, and then keeps on for more hours, wondering when the sun will set ... but it doesn't. What shows as days goes by, and he sleeps by anchoring the vessel on the ice with the hooks that emerge when he presses the lever on his left showing an odd hook symbol. They grip well, first dragging the ice runner to a halt, and then digging into the ice by a mechanism he can only guess at for now. The craft shudders as the wind bears down and then tears upward heavily, but setting the vertical propulsion vanes by pulling the lever bearing a feather above it sets them to automatically adjust to the sudden direction changes coming at it while letting most of that wind through. Three triangulated runners are held by on thin, ultra-strong arms which he is amazed can hold the vessel on course in flight, while now, in combination with deftly embedded hooks, hold it in place in a rhythm automated by the wind changes, so that in a matter of tiny instants they adjust dozens of times, gripping and releasing, without giving way. He can feel it, but as yet he cannot quite understand just how it is happening.

More days go by, or what used to be days, according to those hands, while he fights a kind of nervous boredom that wants him to rest while moving - a fatal alluring he finds annoying and dangerous as he senses those ever lowering lids that must of absolute necessity remain fully open. By repeated experience and effort he has learned to trigger the response in order not to drift too far into a dream, and he has learned this by being frightened during moments when the spell sends his mind away. The craft forgives, but only for the briefest time, and the sense of its drifting sends, by some brain within it always running, a sharp shift in the seat he straps himself to. Only once did it not suffice to get him back, and in that instance, the sliding created a roar from the blades that awakened him to a fright he would not soon forget. Remembering it a bit too clearly now, the craft was raised up on two of its runners by the wind and was about to tip over, when he wrenched the controls and took charge, adrenaline gushing.

"That was the first and last time that'll happen," he tells himself over and over. And so far, he has kept his oath.

Of course he goes through many thoughts as he pilots his way, and he's uncertain of his now virtually useless name, so he never bothers with it, although he begins thinking about the origin of things such as how he came to have certain recollections, of how beings like him would have things called mannerisms, postures, and other modes of non-verbal expression used to identify characteristics, how certain clouds of ice crystals elicit a sense of different emotions, memories, and considerations, but only vaguely. A rambling mind hunting for internal answers, with little else to do except to keep repeating a guard not to crash, can go quite far from the ordinary. In order to keep from brain-dead boredom and going stupid, which he knows he could do, he lets it go to see where it would take him. There was a point when he is watching himself sitting there when he begins to think about how it is that the world is speeding by while he is just sitting there, and how strange such things as relativity exist at all - probing for moments that entrain each other into deep, profound thoughts he feels he would never go into at all in his normal life. Even when considering that, he begins to wonder what "normal" really means, and if there really is such a thing when considering the human experience.

He then notices, has the idea of taking his watching to another level where he puts more intense effort at seeing himself more clearly, and in beginning to do so, he discovers that it requires a level of concentration which is quite difficult while steering the craft. "It's kind of like chewing my food, patting myself on the head, and twisting my foot rhythmically all at the same time - a lot of focus," he thinks, just as he releases his grip on the wheel slightly with his left hand, causing the craft to drift a bit too far from his control, and a wave of adrenaline shoots through him as the beginning of panic begins to make its way into his mind.

This is going to take some getting used to, but he has nothing but time on his hands, and he starts to notice a certain sense that he finds compelling, even if it is at times alarming or embarrassing, revealing things he's never considered before about how he expresses thoughts and emotions physically, in expressions, and slight movements he has never really seen before. It's as though for the first time he begins to see the body he inhabits as someone he can watch. He feels himself observing himself, from what seems like a location outside and above his own head. Once again it brings out a strong reaction, loss of control, and panic, but he craves for it to happen again, and as he begins to become more adept at the practice, he finds that he can actually see himself piloting this craft as it speeds down the ice at breakneck speed.

Then he sees something out the corner of his eye that he corrects mentally as "just a figment," until he sees it again an hour or so later, and after the third "apparition," he decides that he isn't just imagining it at all, and that one way or another, he'll have to investigate. Hours go by as it appears again and again to his right during moments when the racing and swirling clouds of ice crystals subside, and finally he decides that he has to steer away from the waiting sun long enough to get a clearer view. Within minutes, he also sees that whatever it is, other than that speck he saw originally, it is appearing to be larger than his motion in its direction would allow, and that whatever it is, it's coming towards him just as he is moving towards it.

The shape of another craft like his begins to take form. At first he can't interpret it, much as the natives who saw Spanish ships approaching their island couldn't comprehend what they were seeing before them when Columbus landed in El Salvador. Then it strikes him, and there it is, sailing along the ice. As they approach each other, he can hear, despite the howling wind, a rattling that is not ordinary in his mind as the signature sound of an able craft, and as he gets closer, he sees the other pilot signaling him to stop, as she does also.

This requires a degree of expertise he needs to muster from somewhere inside because he's never done it before, and it requires that he take much more care to keep his craft from running into the other and potentially killing both of them, because although the hooks are very good at braking, they weren't as easily controllable as he imagines. Thus the stopping takes time and patience which the other pilot finds difficult to manage, as he can see by the waving and apparent shouting from the other craft. But it is accomplished, and he waits for the pilot to exit that craft, only to find that exhaustion has apparently taken them over after an effort he didn't realize would be so taxing, and he slowly makes his cautious way to the other vessel, discovering that its pilot is a very tired woman who doesn't want to open the small access window to speak to him. So there he is, wind whipping him to and fro as he hangs onto one of the runner legs, trying to instill in her the impression as best he can that he isn't leaving until she speaks to him, and she finally twists the latch lock to slide it back.

He asks if she needs help, as the sounds that are coming from her vessel signal what he has construed as a dangerous loosening of a runner arm. She nods, shouting over the loud winds that she doesn't have the strength left to do it herself, while passing him the few tools needed to tighten things back up. He realizes while doing the job that she has probably been dealing with that arm for quite some time, over and over again, because it takes a great effort for him to bring the arm mount to a tightness he has faith in, so that the situation won't repeat itself soon. Being out there exposed isn't fun, and the wind and ice burn he's already gotten from his own outside efforts of prevention, along with the replete telltale signs on her face that he can see to the same effect, he wants to be as sure as he can that things are as tight as they were going to safely get, as any form of forgiveness of this desolate and dangerous world has not shown itself to be easily had. When he is done, he makes his way back to that small window and asks if her provisions of food and water are OK, and she asks for some water, which means that he'll have to go back to his craft to return with what he can spare. After the retrieval, he hands her a small half-full bottle from among the empties he's collected, with some in it, and she swallows it all with a bit of desperation that also tells him that she is dehydrated.

"I've almost given up," she shouts, pulling half a dry food bar out to gobble down.

He smiles, trying to show sympathy and affirmation. She returns the gesture, asking what is next. Such a thought had not yet dawned on him. He stands there vacant-eyed for a few moments, and decides that they need to get those craft as close together as safely possible, so that they can find another way to communicate.

What seems like a simple suggestion turns into an hour of painfully repetitive and increasingly tedious maneuvering. He knows that his patience will be tested, so attempts to gear up for it, and begins to sort out the little tasks his craft requires to inch next to hers. After repeated attempts at getting his craft to face the same direction as hers, he finally resigns himself to go into the wind so that the two vessels are facing in opposite directions. Although his initial thoughts that it might be impossible to do, he discovers that with a good, slow, sweeping arc around hers, he can shut down the wind vanes on his vessel just in time to slide quite close so that they can shout to each other through their small windows, and then the same thing occurs to both of them, and they discover each other almost simultaneously holding up a self-erasing writing pad they find next to their seats in the cockpit. There they are, sending notes through those scarred side windshields, and to their surprise the communication goes well and quickly. No sentences - just phrases that do well, as they both fill in the blanks as though they already know each other. They decide that they need to get some signals straight so that seeing to each other's needs can be dealt with quickly, that they shouldn't veer too far from each other in order to keep the communication between them steady, that they should search every inch of their vessels' interiors to see what they have available for whatever uses they can come up with, and that they should stay there for the time being until they both feel that they are ready to continue on. Each discovers that the other has been traveling for weeks after a sudden awakening in the world in which they find themselves, and that neither can recall their name or background, although they seem to retrieve skills they have, along with having a strong sense of commitment to keep flying on the ice towards the giant sun ahead, while not knowing exactly why.

Together they have sufficient food for at least a week, but she has run out of water from nervously drinking more than necessary in her first few days, while he has a week's supply, and sees that there is a small heater in each craft for making a liter or so from ice crystals that collect outside in a nooks and crannies on his craft. She then collects her own and melts it until she refills her empties, as he does.

They share food, water, and rest, and then begin preparing a plan to help and protect each other for the inevitable launch, which is much more difficult for him as he needs to turn his vessel into a position where the wind is no longer opposing him without risking crashing into hers. That done, he quickly makes his way next to her at what he estimates to be a safe distance, and they shoot towards that sun again, thinking all sorts of new ideas and strategies. Hours later, she spots a fleck way in the distance and signals for him to follow. As they get closer, they both see that it is a downed craft like theirs that is slowly sinking into the ice, and although it looks to be in better shape than hers, which is now showing signs of wear he's not seen before, they know that it can never be removed from the ice without more help than they can gather. So they circle it, parking and anchoring theirs closely downwind away from the full force of the wind, hoping to inspect the craft and salvage what they are able with the little extra space theirs afford. They find food, water, empty containers, and extra tools, stowing them well, and he begins to remove the mount parts that are in question on her craft so that he can set about repairing hers to full, safe functionality, as it has already vibrated loose its own mount in that time they have been traveling together.

While he is attempting to do the repairs, she is thinking. She begins trying to analyze the ice to see if she can find a way of perhaps building some sort of shelter on it by packing the crystals that gather around anything that isn't flying along with the wind, and notices motion beneath its surface which at first glance hints to her that there might be food there in the form of some sort of fishes or other creatures. She kneels down on both knees and rubs a small arc so that she can see deeper, as it is covered with a thin layer of crystals that prevent a clear view, and when she does, what she sees is shocking. Those are not fish at all, or any other underwater creatures, but strangely, and almost unbelievably, life - life that is not aquatic, but somehow, in some form of suspension, embedded in the frozen ocean. Life that has people, plants, ground, buildings, artificial lighting - all fully functioning as they would anywhere on Earth. As she bends down and gets a much clearer, more detailed view, it is as though she is looking into another dimension - "through the looking glass," she thinks instantly, as she looks at more and more detail below.

A small group of people are sitting around a table eating, while in the midst of a heated discussion which when she pays more attention, she senses that she understands what is being said. It appears to be a family argument over the leaving of one of the two children, a teenage girl, who is trying to convince her parents that she is old enough to start going on her own. "It's English!" she says to herself in disbelief. Yet it really is, and not only that, but it strangely reminds her of something in her past which she cannot remember well - a sort of déjà-vu experience she can't quite grasp, as her memory of her own history is only there in blurred fragments at best. The emotions of the moments she sees below begin flooding into her mind, as if in a swoon, and suddenly she feels a pulling both from below and above, as she then hears a familiar voice shouting at her from a distance. He has grabbed the hood and shoulder of her coat tightly and is pulling with everything he has, as he loudly admonishes her, over the howling wind, to pull herself together and try to pry herself free. "Free from what?" she asks herself as he continues to pull, until she notices beneath her that both knees have started to sink into the ice, and so have the toes of her boots. She begins to struggle, placing her hands next to her knees and pushing as hard as she can in order to free her knees, which are now in the ice deep enough to keep her from getting up at all. They slowly begin to pry upwards as she fights nervously and frantically to get out while he continues to pull. Finally she feels the ice give way as her knees break free.

Once out of her predicament, she manages to get into a semi standing position as he grabs her waist and continues to pull, while shouting to her.

"You were almost stuck completely!" he rants nervously, as she shakes her head in disbelief, tossing aside a bit of what feels like grogginess, trying to sort out what has actually happened. Any explanation at this point is inadequate, so they both go about ensuring their survival by gathering everything together and returning to their craft. Hers is now near its original from his hard work.

The pilot has apparently been dead for days when they arrive and has dehydrated in the cold, dry air, and although they feel they should give him a proper "burial" of some sort, the wind and cold have their own ideas about that, so they make a short prayer for him before they resume their quest, leaving the craft-turned-coffin to slowly continue sinking into the ice. They try to determine what killed that pilot, and eventually conclude that he had simply given up, being alone, since he still had provisions which he may have simply stopped consuming out of destitution. At that point, it is becoming eminently clear that alone neither of them can last, but together, they may have a chance - at what exactly they have not much of a clue. But they aren't going to give up just yet.

Days later, and several sinking craft as well, they come upon yet another with another woman pilot who saw her final moments in the midst of the ice, her craft now sunken well above its blades. The work he did on hers did set right and there is no more trouble, they have enough spares, probability is on their side about the food and water situation; the number of stopped craft they have already found leads them to think there will be more ahead.

After a brief inspection of the sinking craft's storage holds little of use, he starts to shut the hatch when he hears what he could swear is a whimper from beneath a sweater there, despite his added feeling that it is yet another impossibility. Lifting the garment cautiously, he finds a small furry animal curled up and shaking from fear and cold. When it sees him, it has already decided that it is going to die and begins to drop into the sort of stupor an attacked prey will when faced with the inevitable from a predator, and of course when he reaches down and picks it up, wrapping it in the sweater, it immediately begins to make sounds which he can only decipher as joy and gratitude. It is definitely not anything he's ever seen before, a combination of miniature dog and monkey assembled into an appealing little body that appears both harmless and cuddly, and this is precisely its character when they get to know it better. It first becomes her passenger, and then alternately his, as they trade back and forth at stops, after feeding it food and water until it doesn't want any more.

They resume their trek to the unknown destination, toward the great sun that they could swear is getting ever so slightly larger the more they speed towards it. The little mascot has shown itself to be smart and willing, so they no longer restrain it, and now it sets itself poised with its front feet at the windscreen, sitting on the lap of whomever it rides with, eating and drinking little enough not to be considered at all a burden, while providing body language and sounds that, although unfamiliar to them, are readily accepted for warnings and interest at whatever it senses or sees.

The three become a team that works well together, until one day the furry creature lets it be known that it prefers her company. As it becomes his turn to take the little passenger, he goes to lift it for the switch to his craft when it darts away, and he lets it stay without argument, despite a slight pang of rejection mixed with a little jealousy. They speed onwards, determined to find something, but there is still no sign other than the very slight commonly construed increase in the sun's size as subjective evidence.

Without the presence of the little creature to occupy him, he reverts to self-observation, at times even forgetting the presence of the other craft. He notices that his craft has veered away from hers, so far that he cannot distinguish her face. Later, when he asks her about it, she responds that he is weaving very slowly back and forth and that she is keeping watch to make sure that he doesn't get too far away. She isn't worried about it, he concludes, because she has her little friend as company, leaving with a hollow feeling of loneliness after all these days of being alone.

It is as though there are the two of them, and then there he is, alone, but with them. His gift of imagination begins to run with such ideas, and this makes what was once a clan of three into two plus one. This in turn makes him keep more to himself, and it then builds into a slight feeling of animosity that shows at times in their conversations, which begins to annoy her, so she offers to give him the little one as company, but he refuses.

And so it goes until they run out of food and become worried, they stop at another sinking vessel only to find it abandoned, and no food. The frustration reaches into each of them and an argument begins because he feels that since the little one has become the designated food sentry as they speed along, that it is its fault that they have to go through all the motions only to find nothing to eat. She doesn't agree, and comments about what a narrow remark it is considering everything, and then they give each other the first look of distrust.

They both jump into their crafts muttering to themselves, and then jump back out, shouting that perhaps it is time to separate, since they really don't need each other that much, amidst the rumble of their stomachs growling, and the little one hides in her storage, shaking. It has the feeling that something is going to happen that it doesn't want to, but all it can do with the feeling is quiver and feel confused, so it bundles up tighter and waits with its eyes closed. This looks as though it can be the end of the three of them, as hunger and survival instincts begin to make each want to either take control or go off on their own out of growing desperation. As though they have already agreed to do so, they both return again to their craft and began to prepare to separate for what might become a final time, when as they look ahead, each sees something in the distance that makes them hurry to launch. It can be one of those thoughts each has had that it isn't really anything, but when they both can sense in the other that slight urgency, it compels them to jump at the opportunity at least to see what is ahead, and they then do just that with a purpose driven by the present intense instinct of hunger that they both are now feeling.

As they approach they realize that it is yet another craft like theirs, with a withering pilot's body inside that is beginning to shrivel from dehydration, and when they park their vessels and see that the storage has no food, they become even desperate, looking at each other with a glance of hopelessness, realizing that they are now stuck with no way out, and growing more hungry by the second, as they haven't eaten in three days. They stand there trying to think, the wind howling around them, the body of the dead pilot before them. She turns to him and says in a gutteral voice, "You know what we have to do."

Yes, he knows. It is to either start cutting some drying flesh from the corpse in front of them or to risk death. There isn't much else, except getting back to hunting in the open vastness, and both of them are so hungry and worn from lack of food that it only takes a few moments before he gets the knife from his utility kit and cuts into the jacket, sweater, and then the arm of the dead pilot's body. Once he has just begun to move towards the corpse, she is already right there with him. Of course they are worried that the pilot might have died from disease, but it really doesn't matter that much at this point. They know that the small heaters which they use to melt ice crystals for water can also boil it, so he cuts out a sizeable chunk of the pilot's bicep and then begins cutting it into small cubes as she rushes to get their heaters. They are both well practiced at getting their craft together with each other and moving between them by now, so he isn't worried about her more fragile frame being tossed away anymore, and he remains intent on cutting each of them enough to fill both vessels as she returns with the heaters.

"Fill them half-way and boil the water," he says so quickly that it startles him, and she moves to do it just as quickly, as they begin to salivate from the mere thought of food coming soon. What has not that long ago been sufficient grounds for them to separate, perhaps even permanently, is now long gone, and the smell of the boiling meat, human or not, leaves any inhibitions they might have about doing what they are now long behind as well. It is food - most likely good protein that they need, and that is all that matters right now. They eat, and then quickly get to boiling some more, while handing bits to the little one as they eat. Being much more focused on the present and its implications than he, she lets him feed it as she continues with her meal, and the little one instinctively does not forget who is feeding it, as it tries to move towards the one with that food for the first time in quite a while. They are three again, brought together by hunger and cannibalism. Whatever rules that might lie remain the foggy back of the two minds has lost the battle, and after eating their fill, they cut away more human flesh to take with them. Something in them makes them turn to the pilot's body and imagine his spirit still there, however, and with clarity of renewed human gratitude, like primitive tribesmen they pay homage to the source of their salvation before them. Yes, they are grateful, and yes, something in them wants to express it before they return to the journey ahead, satisfied now that although everything isn't back to normal, and far from it, they are able to continue nourished and ready for whatever is coming.

She has cut a lock of hair from that dead pilot's body and stuck it in a jacket pocket as a form of tribute. Yes, this is an extraordinary journey she intends never to forget, and she chooses to live with the choices they make, without a remorse born of social inhibitions she can only sense now distantly as brief hesitation. So it is with a primal instinct.

What they have just done to survive proves itself in the days ahead, as their shared, remorseless actions are followed by the virtual depletion of what they took before they again find more sustenance. Late in their sojourn on the fourth day, they finally come upon a piloted craft deeply sunken in the great frozen, and with effort they wrest free a surprising amount of food and water from it. It looks as though its pilot took early precautions on running out of food and water - something they could have done themselves had it occurred to either. But now is another day, and now they set away with at least a week's supplies, free to be in search of more.

But what they come upon instead is another extraordinary vision of flashes of light that keep changing patterns and intensity. As they draw closer, they are once again awestruck, by the vision of a small abandoned city, seeming to be built entirely of ice, or glass, and when they finally find a sheltering juxtaposition amongst the structures and leave their ice runners, the first thing they do is test the material in order to determine its nature, which is indeed that of a very tough glass which appears to be virtually indestructible. After hammering on corners of walls until the impacts take a resounding toll on their limbs, they are satisfied that it is, and looking upon the buildings, the positioning of the sheets of the material implies a fusion into formations much akin to a house of cards. These, however, aren't about to collapse from the high winds. Nothing there budges or flexes as they walk cautiously amongst the crystal buildings, looking for anything else, of which they find nothing. There are no artifacts of any kind, and the buildings are totally empty from what they can see except for the collection of ice crystals that are blown into them by those winds. As they wander from doorway to doorway, there are almost imperceptible fragrances that catch their attention briefly.

"Do you smell something?"

"Yes. It's sort of a flowery smell, but I can't make out what it is."

"I don't smell that. What I smell is like fresh fish."

And then again there is nothing.

"I don't understand. There's nothing here, so how could we be smelling flowers and fish?"

"I have no idea."

And as they continue walking and searching, now even more cautiously and intently, hints of other odors catch them by diminishing surprise.

"There has to be some place here where that's coming from."

But as they continue, they see nothing else but the buildings, ice crystals gathered in corners which they try sniffing at to see if they are the source, and the sunlight that fills everything with a daylight that, where reflections from the buildings gather, is intensified and thus has a warming effect. Still there is no evidence that anything else is anywhere to be found to match to the smells, which are at times sweet, at other times wooden, or like a tobacco, or sour, or acrid, even to the point of being mildly irritating, and what seems like everything in between as well, often mixing with each other into fragrances and odors that can only be described as outright strange to them. Some even feel warm, while others cold. It is at times unsettling, as they can not decipher the sources.

The little one, which she had carries with her to sense anything as well, is also responding to them strangely, as it appears to respond to them more like other animals than mere smells - as though various animals are approaching it - some friendly, while others frightening to it. Then as they find a sun warmed spot to rest on the perfectly polished and precisely flat floor in a room within the interior of one building far from the wind, it tries to jump up and run towards what it smells as though it is from one of its own kind, shrilly whining like a puppy sensing its mother, and then suddenly darting away from another as though it is a predator ready to descend upon it.

Then there was again nothing, and they decide that it is time to sleep at least a bit. His watch has been rendered useless from accidentally smashing it against her craft during the daily checks they make to ensure that no high speed problems will occur, so any reference to what was once time as they had lived with it is gone. They sleep when they feel the need, and this is one of those times. Covering their heads with their spare sweaters to block out the sunlight, they fall into a sleep only warm still air could bring, and their dreams are rich and unfolding into a deep kaleidoscope of new images. Then suddenly they share dreams of a clash between beautiful creatures of which they have no memories with what seems like monsters pursuing and then suddenly turning and running from them, as they then become outnumbered by the beautiful, and then they both awaken, startled by a connection that strikes them as soon as they open their eyes.

Looking at each other and the little one, it suddenly dawns on them that what they are smelling isn't just evaporations in the air, but things much more substantial and, dare they believe it, alive.

When the thoughts come together, the smells become more pronounced and seem to gather around them as though they are being attracted by their thoughts, and they then begin to feel emotions coming from them of compassion, arrogance, hatred, attraction, warmth, repulsion, fear, pride, and many others, each seemingly connected to its own odor or fragrance as it drifts through their nostrils like a crowd of people passing one-by-one before their eyes. Only there is nothing to see, except perhaps brief internal images that arise with the sensed smells.

The sense of conflict again grows, and they realize that this is a real conflict they are witnessing between living things of an order they have not even conceived of before. And what makes it more unsettling is that these invisible beings can elicit emotional changes in them and the little one as well. Not all of them are hostile, as some feel as though they are indeed friendly and even healing, if one could so describe them, while others are unquestionably antagonistic, as though their presence is felt by them to be unwanted - an invasion of their territory. It is then that an understanding of it all suddenly pops into their minds. The air in this room is still, and the fragrances, smells, odors, or whatever they choose to relate to them as, are able to move freely in the still air because it is just that. The city has become a shelter for them from the dangerous winds that would otherwise tear such delicate clouds of fragrance asunder, destroying them mercilessly, and the struggle being witnessed by the travelers is one involving this city's safest rooms and buildings determined by the stillness of the air within them.

The problem seems to be that with such a variety of them, they have not yet resolved their differences, and may not ever resolve them, because without still air, the mixing of various smells is construed to them as a form of integration or racial mixing which each reacts towards as different as they are from each other. The conflict includes anything that would disrupt the stillness of the air in the room as well, and they suddenly realize that it is their own breathing which has put some of these invisible beings at odds with them.

This only makes them breathe more rapidly and deeply, and intensifies the situation. And as they begin to move about, they realize by the reactions they sense that their motion is disturbing, which in turn brings the invisible to elicit unpleasant sensations in them that become greater the more they move about.

It is then she realizes that despite the delicate composition of their gaseous bodies, these living things have real power to elicit emotions in her that she cannot control, and that those emotions run so deeply that they have the potential to injure, or possibly even kill her. And just as she has realized this she turns to him to warn him, and sees a look of pure terror on his face. She grabs him by his jacket and pulls downward, locking her eyes on his, and telling him that he must slowly go to the floor. He is near delirium, but struggling to keep his nerve, and does as she has instructed. Gladly, the mere few words of direction she offers him give him that sense of solidarity he needs to take greater hold of himself.

Each on their knees, she also begins to feel something strange inside, and then tells him to lie down and place his sweater over his face, as she does the same. Lying on their stomachs now, she tries to explain as best she can why, and he calms further, beginning to understand that his emotional situation has been imposed upon him by whatever of those invisible clouds he has breathed, and that hopefully the sweaters would help to block any other entry, at least for the time being. She then turns and, holding the little one, who is shivering for dear life, begins crawling slowly towards one of the doors, and he follows, hugging the floor. They make it outside the room, followed by aromas, but only a few which seem not to be trying to invade, but assist, as all three quickly find themselves calmed, and they continue crawling until they are well clear of the room and begin to feel cold air ahead.

They stand after what seems like a very long time, but which is in reality only a minute or so, and gather their composure and begin assessing the situation more clearly.

"You know what we're dealing with here," she says.

"Uhuh," he responds, wiping his torso as though they may have clung to it as well. "They're alive and dangerous, and we can't stay here long."

She replies, "Well, as long as we stay in air that's moving, and the faster the better. If you're wondering how they could have gotten here, it dawned on me that they could have been here for a long time - and come up from below," as she points downward to show him that the buildings don't seem to be sitting on ice, but go downward as far as she can see.

"I saw that while crawling out of there."

When he looks, he sees that there were different levels below that didn't seem to end anywhere, and he says, "That definitely isn't ice down there that I can see - just more of this stuff. As he looks more, he adds, "and it looks as though where we're at is the top of a huge city, much bigger that what's above the ice."

"I think we'll be safe here for getting things together. At least we're safe as long as the wind helps us," she says.

They sleep there, but the cold wind which protects them also chills them prematurely, so they decide to seek out more warmth, and then sleep more deeply, occasionally popping their eyes open and sniffing, just to make sure they don't smell anything "funny." The little one hugs his neck as though he is now its daddy, letting out occasional tiny moans of relief, and that to adds to their general comfort. However long they sleep, when they awaken, they feel rested and ready to gather themselves up and into their runners again.

It is a funny set of feelings they have as they leave the city, thinking about everything, wondering what else they may encounter, fearing the unknown in a way they have not felt there before, and wondering even more intently what kind of world they have found themselves in. It is a place that now feels much more unknown than previously, as they have adapted to things before, while now their experience has left them with a sense of the unknown ahead that they hold with much more fearful respect, even after having overcome the wind's brutal treachery with such studied poise. Their assuredness has now taken a few steps backward to accommodate unknowns they are less certain of, and although their level of uncertainty has risen, so has their trust in each other. They are closer now than before, and the feeling is more welcome than when they first met, as then they didn't know each other as well as now. Things are different now and they are much more of a team. But then coming out together on the other side of adversity does that, and they are becoming more savvy to such things moment by moment.

As he is steering his way down the ice, feeling almost normal, he allows himself briefly to recollect the range of emotions he has been through there, and as he does, they rush and well up within him all too vividly.

"She's there now, and I can rely on her ... and the little one," he thinks out loud, encouraging himself in the face of the fears he is feeling that are too fresh to dispel. For the first time, the loneliness he felt so long ago begins to come back with a new face to it. He now has much greater feelings for them, while at the same time, the thought makes him feel an unfamiliar twinge, as the fear of losing them on this quest, for whatever reason, becomes more vivid. Wishing that he'd never entered that particular thought, he brushes it aside, but it lingers and slowly trails behind him, then is caught in the wind that destroys most things in which they are caught, while it drives them ever forward as well, cleaning the ice before them of obstruction. It tosses crystals that flash as they are driven along, often filled with the light of a sun omnipresent in a world of wonder and awe of which they have yet to learn all mysteries, and have solved even fewer.

With craft loaded with supplies and a sleep that has helped them recover from their recent encounters with things which although scrutinized well for those moments, are yet far from fully understood, occasional reassuring glances between them have all but replaced their need for writing on tablets for communication. They become more and more of a single mind as they fly down the ice towards a farther horizon filled with the unknown. Their thoughts hold a wonder of what will come next, and when it does, will they be prepared for all it may involve? For a moment, there is a glimpse of the memory of the world below them where involvements vast and virtually endless make them hold on tighter to their wheels, as though the only thing which can really save them, they can imagine, is their own abilities as yet not completely understood, and the dauntless craft that have gotten them this far so well.

Hours later, according to a time whose divisions have long since been lost, they sleep again, and their dreams are filled with stories that could be from that ice below, while the little one in troubled sleep emits tiny whimpers and moans decipherable only to itself, as it awakens at moments to hear his breathing, and the sound of its slow rhythm brings it comfort and more sleepiness to return to. They know so little of each other, yet so much at the same time, while the world outside blows ice cold danger, and an almost certain promise of experiences so very few may actually have been resolved, if any have at all. Only that unaccountable time they now live in will tell, but it is not speaking as yet of what it will bring.

In the midst of their slumber-born visions, a conversation arises in the background of decisions, questions, possibilities, and conclusions that they cannot decipher, the currency of which they presently cannot hear or interpret if they could, and the depth of which even their imaginations cannot contain. For the vast and changing globe upon which they rest is so ordered that neither their place nor their time could hold all of it. Still unknown reparations have yet to be made, and their only way of gaining them is through the knowledge of experiences yet to be had. Their journey is about to unfold, and their sleep is but a respite for actions they must continue at a pace they cannot know until their present moment demands it. But now they are refilling their reserves for this, whether realized or not, and that array of future circumstances which will make itself known awaits in the eternal present of another day coming.

As they awaken, pulling those fabrics from their heads, a relief sets into their thinking as they look to the horizon afresh with clearer, rested minds. They take their food and drink thankfully, and think of all things necessary before they must embark again, while the little one yawns and smiles, also looking ready for what the day will bring. What at times was hurried now takes a slower, more experienced pace as they go over their craft and ensure that everything that is meant to keep and support them is ready and willing. They have now gotten to the point where the routine requires much less sorting, as they have developed their patterns well, and although the speed of their chosen movements is lessened, the desired result of their already well considered machinations is not.

Thus they speed away from their rest endowed with an expertise born from what was given them mentally and what they assembled themselves along the way, now more resolute in their thoughts, armed with more patience as well. And their expectations, although sharpened by their recent past, are more considered, as the shared stillness of their subconscious mind has done its own sorting of experiences in a framework of a dexterity and poise elicited by the needs of each moment as they arise.

Two sleep stops later at about mid-sojourn to the next, there is again something in the distance that catches their eye, although the little one doesn't seem to be responding to it. Perhaps it is one of the mirages they have learned to detect, but their sense of it hasn't triggered the bell yet in their thinking, so they keep onward towards it, slightly off their normal course towards their source of all light, and although they have traveled for some time in that direction, it doesn't seem to be getting much larger. In their interpretations and imagined ideas, either it is thus very far away, or it is moving away from them as they approach it. Keeping on, they can still not seem to get much closer, and they are becoming tired and hungry. So after an hour's farther run, they decide to eat and then sleep again, deciding that upon their next awakening, they should continue onward no matter how long it takes as long as they don't feel any danger from tiredness on the way.

They have by now become so well versed in parking their craft that shouting conversations are possible from the craft's small side windows. Their conference on the next goal is thick with talk on how to detect and avoid dangers much more cautiously, and what to do if there is a need by brief unobtrusive hand signals that are well tested there beforehand. They aren't going to take any chances this time, and they try to cover every possibility, despite not knowing what to expect. They have a few days' food left, and she rigs a sort of pack for the little one to perch in that will hang on her chest, so that she can move quickly if necessary, while being able to sling it around her to the back in one motion so that if full-tilt flight on foot is demanded, she will be ready as well. They talk until they feel it was useless to continue, having exhausted every conceivable contingent, and then set themselves in comfortable positions and sleep.

Their dreams are once again, as has been since their escape from the city of glass, filled with images and emotions that have grown progressively dimmer, but are still lingering. There are still things in their subconscious that seem to need resolving and sorting, and the dialogues are once again there in the background of their dreaming, still nearly imperceptible, but growing less so. Words emerge from time to time, and although their sleep is deep, a sense of their own presence in them begins to grow.

They see themselves at times in them involved again in scenes reminiscent of those locked into what is beneath them, and the familiarity sensed as their eyes rapidly roll about leave some impressions which stay even when they were fully awake, popping into their minds as they prepare for the long haul coming towards whatever it is in the distance. They toss about the reasons for its seeming inapproachability, first picking one reason, then the other, unable to decide, while in the distance looms something they know almost nothing of, waiting silently. They have even left the possibility for other reasons for its distance, while attempting to reach an alternative is impossible. So they board their ice runners and break loose, now storming towards their now sitting target, waiting in its own sense of silence for them to draw nearer and nearer.

As they speed along, they start to see the dot turn into a much larger object, and then as it grows into something clearer, they start to wonder if it really is what their eyes are telling them, as their previous possible conclusions about it have had to change rapidly. It takes only a few hours to approach what seemed to take much more time before their last sleep, and it is growing rapidly before them, growing ever larger as they approach. There is no mistake anymore about what they see is ahead, as the mountain before them takes up its portion of the sky they have become so accustomed to being immersed in, and yet grows monstrously larger as they near its huge footing on the ice.

With the rocketing air currents picking up as well, they suddenly decide that they cannot get to it in the position they are in relative to the winds, because they feel their vessels skidding sideways on their blades. So they bear a hard left and try to directly ride with the much higher wind being deflected by the huge mass, and follow it around its base, letting that wind choose their path rather than fighting it in most dangerous futility. It works, and as they are taken around the mountain, the wind is then continuing on its original path, while they use the opportunity to tack towards the slower side of it that is forming to their right. In doing so, they discover that the air currents are dying down considerably, and they hope that there will be a length along the base where it will drop to the point where they can coast more slowly to a stop close to it. After a time, it becomes so. The craft finally cut loose from most of the wind's power and do indeed coast towards and into a small cove that looks to be waiting for their arrival. The heartening relief they feel after the sudden struggle with the forces of nature gripping them so strongly did well to put them in fear for moments that seemed to last too long for their nerves. But they are now settled in and prepared to explore the giant in front of them, making sure that their preparations for whatever is ahead are as sufficient as they can make them.

They exit their craft, confer a bit on their approach, and then begin towards as easy a footing at the base as they can quickly find. It is steep going as they begin to climb, and they are not used to dealing with the grade, so they have to rest often, but they are resolved to keep at it. The ground beneath them is quite unfamiliar and they can not assess its composition. It feels like something more than just frozen Earth, as there is an indescribable and unsettling quality about its variegated color that makes them stop, but the grade is so steep that the first thing on their minds is finding a section on which they can securely stand, or even sit if possible, out of the winds that are sweeping around everything in their sight. The ice crystals that collect on so many edges underfoot and around them give the whole place an eeriness, and they can feel the hair on their necks standing and their ears nervously perking up as they search for more solid ground for a rest.

They haven't expected the path that they now come upon, and its welcome sight gives them pause to follow it visually to gauge its navigability for as much distance as they can clearly see. It looks as though it has been well maintained and well leveled for anyone who might find themselves in such a place, although they can not deduce who that would be, as they have already readily assumed that they are the only people on the mountain. Perhaps they are wrong, or perhaps it is not made for people at all, but some other living things which they had not encountered yet.

They will have to see about it, and as they walk up its well-angled slope, they find that their pace has now multiplied, and also that the path is a very long one if they are to use it to get to the top, which they are hoping to at least approach. But as they continue, they find that its undulating surface does in fact create footing problems, especially from the ever increasing wind currents and collection of ice crystals, so they decide to stop and decide how they are going to continue, since the mountain is growing smaller towards its top, despite the fact that they estimate that they aren't at even at one quarter of its total height, and they decide that fighting the ever increasing wind as they climb isn't worth the risk.

He thinks that it might be better trying to find the next level of the path by climbing as directly upward as they can after back tracking several hundred yards out of the wind's greater pushes and pulls. After quite a struggle making his attempt, he finds it too much for what is left of his unexercised strength to deal with, so they are then left standing and wishing they'd though of all this before they started their climb. The wind is quite low in a narrow nook they find from it, however, and the sunlight is creeping over the shadows, so that they can feel a bit of greater warmth for the time present. But she becomes restless, not having tried to scale upwards as he did, and decides to continue around a jutting section to see how bad the winds are on the other side. As she leaves him sitting there chewing on a bit of food, he is uncertain whether he should let the two of them out of his sight at all, and when they disappear behind an outcropping that bends the path, it doesn't take him long to begin missing their presence, so he decides to follow and see for himself.

When he turns the corner that has obstructed his view of them, however, they are not there, and another large burst of unwanted adrenaline shoots through him like a bolt of lightning. He begins to lope along the path nervously, eyes jumping everywhere, trying to see them ahead, but the wind is getting virtually beyond his ability, and he stops, frozen for a moment, perplexed, and frightened. Perhaps they have fallen – he looks downward for them, but finds nothing there as well. Panic creeps into his mind, and he begins shouting for her with no response. He continues, but it is useless. They have simply disappeared, and he cannot decide what to do.

For the first time on this journey, he feels a longing for them that will not leave, and thinks then and there that he cannot leave without them – he would rather die than do so. Out of desperation, he begins looking everywhere, regardless if it makes any sense for him to look, and it is then that he becomes caught by the form of a giant spider on the slope above him. He can swear that the thing is staring right at him, and it is much bigger than he is, so he remains frozen and spellbound by it. And next to it she stands, looking back towards him with eyes held fixed upon him and wide open, with the little one in its pouch on her chest, both immobile as statues, and obviously even more frightened than he is.

It is then the he notices that although the giant before him is there, it also isn't, or so his awestruck mind imagines for the moment, at least. It seems to be vibrating, like a shiver over its entire body, but at such a rate that a sort of high-pitched hum is emanating from it. They now feel that the strange creature is someway speaking to them, but with nothing visible other than its vibration and high hum coming from it. They stand awestricken as it communicates, in total disbelief of everything that is happening. As it does so, it suddenly darts to and fro in a very short time, using such speed that it is only a blur that they see, landing next to him, and then jumping in another blur back to where it was standing, a short jump from her holding the little one. Its lightning nimbleness frightens them even more. They cannot escape from this thing, and they and it know this.

Its message is one of explanation, however, without a threatening of any kind.

"I am the guardian. I am not here to harm you, but to protect those who have come before who gave their lives, and who thus helped you survive."

It gives an invisible hint to them to look closely at the ground, and when they do, it is as though there is a kind of masking haze that has been covering it from their eyes that is now gone, revealing in high detail its composition of the bodies of other pilots, most wearing the same suits they found themselves in on first awakening on the planet, on the same journey as they. There are obviously millions of them, because it appears that the mountain upon which they stand is made entirely of their bodies, somehow solidified into a singular mass. As they look they see beneath holes cut or torn in their clothing that there is flesh missing from every one that they can distinguish. And then they remember what they had done earlier in order to survive. It strikes both of them, and when they look up at the creature, it is already answering their question.

"Yes, each unknowingly gave of themselves after they had given up, and so they helped those who continued after them in a moment of need. They are thus each known as one who sustains others on the quest that you have taken as well, and they are waiting to be honored for their sacrifice, as you shall if your end is the same as theirs. If you are, you shall as well be here, and honored as well."

This is a frightening thought that they are not ready to see as even remotely acceptable, and they look at each other to verify with each other what they are thinking.

It goes on. "There is also no blame in what you did, and in doing it you have honored them as well. You also showed respect for their gift, and this was of intrinsic importance."

A wave of relief that seems emanating from the spider descends upon them rapidly, and it is then that they fully realize that this being is much more akin to them all, including the little one, than they imagined less than a minute earlier, and from their gratitude issues a sense of unity with the guardian. For a moment, they begin to understand the sense and aim of the Being of all beings through him, and then it is gone from them in the next instant.

"You can see for a moment, but you cannot dwell there because you must achieve much more before you will come to truly understand, and in your journey, you will find the reason for your singular aspect of being as well. It is not inconceivable - it is only that you do not have the knowledge to extract the truth yet. But it will come, if you survive. I cannot remain here. I have much to do and you are not alone on this mountain, but once again to be totally clear, you cannot remain either, as it is not your destiny to do so. This is but a fragment of it which you have had to experience."

Sensing something within them, he continues. "Do not fear for me or perhaps the lonely life you believe I may live, because it is not so. You can only see what you have accomplished as yet. So I must instruct you now to return to your journey, and remember me, as what you see of me is but a tiny manifestation of the noumenon which perplexes even the wise, and has for a very, very long time. I could easily tell you much more now, but you cannot fathom the depths of what I will communicate, and thus it will only confuse you. I do not wish to do anything which can in any way hinder your quest, and so I will not occupy your thoughts so. You are already full."

The guardian lifts her and brings her down to his side, and even the little one is totally composed, without fear from the swift action. It stands there, a frightening shape before them, yet they feel no fear, but only companionship with it. They can see that it is looking directly into their eyes, and it says one last thing.

"You do not know what is before you as you rush across the ice. It is not what it seems."

And with that, in a startling maneuver that leaves them both petrified for an instant, it suddenly grabs them both, piercing just the needle-sharp tips of its huge fangs into their torsos momentarily, and then it shoots upward along the grade above them.

A warm sensation floods through their bodies from the guardian's almost imperceptible bite, and something is changed in their thinking that will never leave them.

For a second, they are blinded by it and everything goes black, and then when their vision returns, the world around them looks the same as it was, only somehow greater, more vivid, and much clearer to them, as though their understanding is amplified, and they see more in everything upon which they cast their gaze. The mammoth spider has by then disappeared into the fog of ice crystals tossed by the winds bearing on them, like a host of thin apparitions stretching across the surface of the mountain. When they look down, they can no longer see what they were easily able to a few minutes ago buried in the mountain's surface. Again, but differently, the three are now again alone, and begin returning back down the path towards their craft.

"It just appeared in front of me, and the next thing I knew, I was up on a small ledge on the grade, standing there next to it. I couldn't move, but I was only a bit frightened, as it made me feel unafraid, and then a few moments later, you came around that outcropping below us. For a moment at first I thought that my life was over and that it was going to attack and devour the little one and me. But then it did something to me that erased the feelings and thoughts, and I was left with a sensation that this was far from the truth, so I began to relax, although I couldn't move, even when you arrived."

"I was stuck-in-place myself," he responds, still feeling a bit nervous about the whole thing, including perked ears and hair standing on the back of his neck.

"I'm getting the feeling that we're not alone and never have been - that there's something else going on here. It first started after it bit us. It's as though our eyes weren't really open to everything that's going on at all, and they still aren't, but they're open more then they were."

She barely let him finish, coming back with, "and we're heading into something - it's as though I'm getting some messages from somewhere, but I feel it more than thinking it. It's just unsettling to have this going on inside, even though I'm perceiving a calm, way down I feel something else – like a lot of activity I can't put my finger on happening in layers I didn't even know I had."

"Best to just allow it to take place," he feels and sees himself saying, with no opinion about it. He has no idea where that last remark came from, and it shakes him, but he's been watching himself when he remembers to ever since he began doing that before they even met. He is getting better at it, and the viewer starts to have a life of its own, arising to see him without any effort by him to get it to do so.

He now has at least two minds operating within – his normal self, and this faculty he's now created that seems to be gaining its own life. Perhaps he is actually going crazy, losing his mind, but it doesn't come across that way. It is more as though he is becoming more sane and responsible for his own actions, but he is wary about describing anything beheld or forming opinions about himself that are more than just seeing, and that is coming from within him as well. Witnessing this in himself, he begins to laugh, and she wonders what it is about, so he briefly describes it to her, while she stands holding the little one with her mouth agape.

"I'm starting to get a feeling like that myself – that there's no longer just me in here." Then they are both laughing, walking down the path, keeping an eye on the ice below for their craft.

Once spotting them, she feels like running to them, full of a new energy she feels coming from somewhere inside, but unable to do anything more than enjoy it. He is just walking, watching himself, seeing his facial expressions change in reaction to her sudden movement towards the craft. They enjoy the moment, but they are also apprehensive about it all, as they have not been the happiest of people during this journey, and the new experience they are in the midst of brings up suspicion about its authenticity, so they find themselves pausing to watch themselves and each other going through it, gathering themselves a little more, and continue, while keeping that one percent uncertainty about it reserved, just in case. The little one is moving about in the pouch on her chest, and is also enjoying the contagious relief they are having away from the dense seriousness of everything – the danger, the adrenaline rushes, the uncertainty, and the wonder. It is there at home with them, and doesn't want to leave the pack as it is now for anything.

Since they are now safe inside their vessels, they decide to sleep if they can so that the next phase, whatever it turns out to be, will see them in full readiness and wide awake. They have seen enough to hold that uncertainty close, not quite trusting what they are now experiencing, as the newness of it isn't quite satisfyingly accepted. They are beginning to realize at this point what the wisdom of firsthand experience is really like, and this makes them want to make haste slowly, even more aware of themselves and their relationship to each other, the little one, and everything around them. They will need it in the coming days, and in a way they expect it to be that way. Real knowledge about themselves is beginning to make it that way. They begin to sleep very quickly, and their sleep is deep.

When they awaken, they eat, making sure that the little one has its fill as well, as it is now a member of their pack as much as they are members of his. When everything is checked carefully and gone through, they head out to the open ice and the fierce winds swelling around the base of the mountain. Without even considering fighting it now, they once again let it carry them, and the speeds they are now achieving are simultaneously both frightening and releasing. The incredibility of it is striking them and fueling a desire to push the limits of themselves and their vessels, while their automatic caution is also holding them from doing anything beyond a safe point of challenge, and the next thing they know they are back in the normal winds after slipping out of those from the base of the mountain when it feels right. Now taking a different tack from their sojourn so far out of their previous path, they set sail with eyes renewed and perceptions expanded well beyond what they were when arriving at that cove in the mountain's base. They are now greater than they were yet they are much more interested in the present moment than thinking about such things, and then they are gone down the ice, racing towards a sun that has waited patiently, almost knowingly, for them.

And there are thoughts as they speed onward.

"I look across the vast frozenness with eyes now open. All obscurity, fear, and doubt flee before me, as my heart senses my awaiting destiny. I move forward, and ..."

The sun is warmer and brighter as their ice runners shoot across the immense wilderness. Time speeds by as the present becomes their reality, and they feel themselves drawing nearer to something that is sensed but not known. They then see a glistening ahead, and speed towards it inquisitively. It becomes larger and larger, and the dancing reflections off it from the sun grow ever brighter.

The winds are changing. They cannot speed down the ice as before. Then they realize that it isn't just the winds. They are slowing because the surface of the ice is no longer its usual glistening hardness, and they realize that the sun is now warming them inside their cockpits as never before. Then they see how it is that there is such a glistening on the ice, as they discover a rippling if the light from the sun from it, and a fear of what is coming before them overcomes them. The ice beneath them is now beginning to melt, and all too soon they will be gliding into a pool of water under which the darkening ice is gradually disappearing completely if they do not act, wisely and quickly. A stillness comes over both of them as they are in serious want of any answers as to what to do about this quickening fate they feel that they cannot now escape. A moment occurs between them when they look at each other now that too soon it could be for the last time.

The sun is not going to be the "great solution" for them, despite what they believed until just a very short time ago. In fact, it is looking all too clearly as though it could be their doom instead of their salvation, and this is forcing them into a silence that is deafening. "What do we do now?" she asks herself, and over on his side is a shrug emerging, as though he's just heard every word she is thinking. There is little time to think this one through, and although stopping is still an option both are considering in order to at least save themselves for the moment, it appears that the ice was going to have something to say about it, as they begin to hear, and feel, slight cracking noises beneath them. Then it hits them hard - they cannot stop either. The weight of the craft can make them sink into the world below, and they will be lost as well. A sense of their imminent doom looms darkly in their shaken minds, and they begin to prepare with increasing resignation for what they now perceive as inescapable.

In a jolt of near panic, sensing something odd in the cabin, a change of some sort, their eyes dart downward in front of them, and catch sight of a new control symbol they haven't seen before that is showing itself on the console of their gliders. It is an airfoil profile just to the left of the vane feathering control.

Perhaps they simply didn't see it before because they had no need for anything else - it was much more reasonable answer than others they were considering. Did it really matter? "What a time to start explaining things to myself!" he blurts, as he reaches for the lever beneath the symbol. She is now right with him, he notices, as he glances at her.

"Yes," she replies mentally, "I'm going to pull this as well. We have no choice, do we?"

And then they do it, in a synchronicity of action borne of a sharp and clear level of contact between them they've never had before. When they do, they are not prepared for what happens next, because towards the nose of each craft a canard quickly opens itself, while wings are also jutting from the sides of their vessels to the rear along the runner arms, and within a few seconds, they are becoming airborne. The wheel of each craft suddenly frees itself from its rigid mount, and they discover that pulling those levers in each of their craft has just turned them into an aircraft they could not have imagined previously in all they have already been through.

They rise up from the melting ice and find winds that were not below, and they began to pick up airspeed that allows them to rise up even farther above the planet's surface, so that the sun appears almost directly in front of them. They absorb the shocking change of circumstances with poise and relish, picking up even more speed as the winds pick up even more and well above the melting ice.

"I guess this is it now," he imagines, looking over at a woman alongside who is already nodding in agreement. They speed towards the sun, picking up airspeed, and with near glee they relax again, while the little one, who was unaware of the details of the transformation, responds to his change of energy by peering out with a prominent smile on its face. She sees it, and gives it a smile and a quick little wave, and the two of them smile at each other from the different craft, as the sun begins brightening even more.

They cannot look directly into it anymore, as there are no longer any filtering clouds of ice crystals being blown in front of them by the winds. It is then that they remember those small singular packages in pockets on the right side of their cockpits, and they grab them, open them, and then quickly stretch the almost opaque shields across their eyes, fixing the bands that hold them onto their heads. It takes almost no time at all for them to adjust, as the shields seem to be varying their filtering, somehow adjusting to the size of their corneas. "There's some very serious science going on here," he thinks, as she is thinking virtually the same thing, and they immediately begin to understand something more about what they are doing there than ever before. As the filters set themselves they start looking directly into the sun, which they couldn't have possibly done before, and as they glide along they look at it with yet another newfound wonder.

Then a jolt of clear, hard truth hits them. They could not have possibly seen or understood it earlier, as it took their own mental changes after the mountain, filters and proximity they are now at to allow them. What stands before them is not a celestial body. There is no longer any doubt whatsoever that it is not a sun at all. It is indeed something entirely different that has been disguised from them by their previous limited abilities, and now they can clearly see that what they have so devotedly sought and struggled towards is not what it has seemed to be at all. Its flatness and distinct parts betray a structure which they could not previously detect until now, and surrounding its edge is what they could clearly distinguish as an opening that precedes it between them and the structure. It is indeed a giant hole they are flying directly into at full speed, with the "sun" sitting well behind it and larger that it is, giving the impression from a distance that it is a solar orb. It is not, and again, adrenaline slams its way through them as their illusions about it all are utterly destroyed by unmitigated reality thrust upon them unawares.

They are now flying directly into a vast unknown, and they have to make an immediate decision about what they are going to do. Even the "sky" they have known and believed to be something they felt as so familiar is now impossible to really identify, but they at least know that it isn't what it has appeared to be all along, and the guardian's words of warning ring in their minds like the bells of judgment day tolling for them.

They look at one another as the little one responds to his sudden shifting and postures with its signature quivering, they touch those points on their bodies where it had bitten them, and then suddenly looked downward in unison.

"It has to be this way," they both think to themselves in finality, as they push forward hard upon the wheels before them. But those wheels refuse to move. They can be turned and lifted, but not forced forward and downward. The Great Loss is now before them and with blistering reflexes their total reactionary response is to hunt in utmost desperation in their cockpits for anything they haven't seen already. Under the levers which have evoked the wings that have just saved them are symbols they once again have not seen even though they could swear that they scoured every inch of their cockpits. They are obvious symbols of waves, and with a crazed abandon reaching almost into frenzy they push downwards on the levers, praying for something of which they have no idea, and then it strikes them. The front canards of their vessels shrink, as do their wings, to less than half-size, and their craft suddenly begin to shoot downward in a high speed free fall, straight towards the water they have just struggled so nervously to avoid.

Believing that the noses of their craft are certainly streamlined enough, they do what they have by now learned to do so well. They put their full faith in their vessels, despite their fears of destruction when they hit the water's surface, and ride it out like ancient warriors jumping into the lightless empty void waiting off one of the dimly clouded high cliffs of Aztlan. The water is coming so fast that they can barely react. The noses of their craft strike it as though they have been maneuvering flawlessly by an unknown brain resting somewhere within the bodies of their strange, dauntless vehicles, and in an instant, they are in blackness...

A deep, dark womb of another existence is at once upon them as they have done the unimaginable once again, going against everything in their old nature, and against their minds' capacities to understand. They have dispensed with all their disbelief, giving way to what they could not possibly comprehend, driven by the expediency of moments which they can barely grasp, let alone confront directly tackle with their minds, reeling from awe, wonder, incomprehensibility. But this is their lot – this is their destiny, and this is their future, and it all really doesn't matter one iota how they may feel about it anymore. There is now only one rule for them, and that rule is acceptance of whatever is being put upon them, and the quicker they find that rapidly changing corner of their minds where it lay, the better their chance of survival, and ultimately, understanding.

And then there is nothing but deep and unfathomable darkness, in a sea of dreams... . . . .

Thus ends Part I of this trilogy.

Editors' Note: The author reluctantly but graciously agreed to minor editing of this chapter.

•••

MICHAEL F. HANNON

Michael F. Hannon is an inventor, artist, musician, and writer. He began classical Bible studies at the age of 10. He is a member of the Order of DeMolay, a Cum Laude Society graduate of Mount Hermon School, Massachusetts, and a Graduate, DeVry Institute.

He is a combat-disabled Vietnam veteran with more than 30 years of study in esoteric disciplines including Mystical Christianity, Gnostic Christianity, Kaballah, Gurdjieff and Sufi studies, Buddhism, and Zen Buddhism.

CHAPTER 19

COMET ATHABASCA AND THE LIFEBOAT GENERATION
2025 TO 2037

DAN SHAW, B.S.
AND
SHERRY BELL, PH.D.

As you know, honey, you don't see any women today with the big pregnant belly that I have in that picture with your Grandpa Hal. You probably haven't even given much thought to how babies are born today... but kids didn't always come from a laboratory!

You know how much I love you and how grateful I am to have a grandchild like you... and you know how I always tell you I'd do anything for you? Well, I really mean it. Because I didn't always have you.

I. ATHABASCA

When some small groups of people prophesied the end of the world in 2000, and again in 2012, we laughed and cried. But on July 23, 2024 scientists across the world stared transfixed in the sky at the newly discovered comet they called Comet ME–2037. The newspaper headlines in every language called it *Athabasca*.

This scientifically validated Doomsday scenario had a far greater effect on the mass consciousness than had global warming, perhaps because global warming had the unfortunate modus operandi of occurring intermittently, in seemingly slow increments, in various guises according to each region. Although international space agencies had been observing comets and ringing alarm bells about the possibility of a massive comet hitting the Earth, until we actually saw one in the sky approaching from a distance of 12 years away, we never really mobilized to be able to protect ourselves. We might have saved the Earth had we heeded the prophecies of the indigenous Nations such as the Maya and Athabasca.

Now that the potential Doomsday was staring us in the face, it seemed that every tradition could indeed point to a prophecy and read into it Comet ME–2037. Even the Pope, Sistine II, publicly announced that it was the known to their concealed traditions as "Wormwood".

In the early years of comet science NASA had, as discoverers often do, badly named their discovery "Near Earth Objects." One could hardly expect governments and societies to get whipped up about "objects" that are only "near" Earth.

An amateur astronomer in a small town in Canada was the first to actually break the story in January of 2024, and then the floodgates of news opened. Several national space agencies, including the United States' NASA, had independently discovered the Comet hurtling directly towards Earth, but they withheld the Doomsday discovery from each other, and from the public, for months. They rationalized that by withholding information they were protecting people from panic. Ironically NASA released their UFO files on the same day as they released the alert about ME– 2037, "Athabasca".

And then it took a full year for the G-30 council of governments to arrange any kind of significant cooperation. It turned out that the United States had two vast underground refuges, one well-known in Colorado, called NORAD, and another secret one in Alabama. A few other countries had smaller underground shelters, but those could only hold tens of thousands of people each.

In this decade-long convulsion of feverish self-preservation, the world's economy was the most robust it had ever been. Countries came together against a common enemy, and no resources were destroyed or squandered in wars and deadly conflicts. The migrations completely dwarfed the previous migrations seen in Dust Bowl I and II.

Although the science of comet impacts was practically non-existent, people seemed to embrace one theory above the rest, that coastal areas were somehow more dangerous. Perhaps it was because of some of the ancient prophecies, or maybe it was just that the idea caught the pubic imagination, or perhaps it was because of a deep inner knowing.

The coastal areas had been emptying already, due to rising sea levels and tsunamis, but now the west coast of the United States was about as depopulated and lawless as it had been in the times when it was first settled. And the west and east coasts of all continents were beginning to be as abandoned. In 2037, most people actually had left the coastal cities, and were living along the migration routes from coastal to inland areas.

In 2025, our family moved from Seattle, United States, to Denver. That includes your grand auntie Sammie and great uncle Ben. That's their picture in the hallway, the one in the plain gold frame. Now let's get a fresh pot of mah jong tea and sit down again, because I have only just set the stage for the story I need to tell you, the story I need – you need to know.

II. THE UNEXPECTED RETURN OF THE STARFISH TACHYON LIFEBOAT NUMBER 4

We call the children born between the years of 2025 and 2033 the Lifeboat generation because it was in 2025 everyone on the planet was focused on four Tachyon Lifeboat Rockets, a 'family' of rockets, certainly, but no family ever looked so little like each other. To maximize survivability and diversity, each of the LifeBoats was uniquely designed, though all of the same modular geometric facets. Though the ships had at first official designations and grandiose names in the native languages of their host countries (English, Sino-Japanese, Russian and Indo-Pakistani), those fell out of use because it was so natural to think of the four as Arch, Horn, Coral, and Starfish.

And although certainly infused with grief, suffering and hardship, these years are perhaps the proudest in our global history, and certainly in our family history. The Klay-Varley extended family at first was disgraced by their near-maniacal advocacy for anti-matter lifeboats, and because of the billion-dollar collapse of the national economies that invested in those first few fruitless years of development. How lucky we are to be alive today and to be remembered instead for our successes with the Klay-Varley Egg Baskets.

Malitsa Klay-Varley will be remembered as the "grandmother of Martian farming" for her invention of the "suspended incubators," which came to be known as "Egg Baskets". These Egg Baskets carried the seeds of plants, and eggs of animals and humans not only to Mars, where they made life possible, but also out into space all directions from our little Earth with its uncertain future. That's Malitsa in the picture of all those machines wearing that long dark coat and gloves.

During the LifeBoat years, it came to be said that "the men looked up, and the women looked down," because we saw our civilization polarize into those who were betting their lives on building the LifeBoats, and those who were building the safe enclaves for Earth. Of course we knew that both sides were needed. And of course there were those totally focused on building the Deflector.

By the year 2035, all of these projects were completed. All four Tachyon ships had launched successfully in 2033. It's hard now to imagine the anguish, terror and awe of seeing basically entire cities board four untested vessel and depart for unknowable destinations, never to return. Nearly everyone on the planet had relatives, or knew someone, who was aboard one of those LifeBoats. Those thousands aboard each of those four, the Arch, Horn, Coral, and Starfish never expected to see Earth or one another ever again. And certainly we have never heard again from the Arch, Horn or Coral. That part of our family, and that part of the Earth, so far as we know, is lost to us forever. We can only pray that they have found survival and success, whichever Door they have passed through.

The Starfish, in the Spring of 2034, reentered Earth orbit to the shock of everyone, particularly me, since I did not expect to ever see it again, and had cried for weeks and months over the separation from my cousin Cindia. Though we were never closer than those many years after her return, when she died I had no more tears left for her going. I supposed that's why I've adopted Mela and Mena, because they are the only two who were on the Starfish with Cindia who now live with us here on Mars Colony. Somehow I think Mela and Mena remind me of our family ties to Earth.

Why don't we stretch our legs a bit before I tell you the piece of our family story about Orbiters and you'll begin to see why all this is so important for me to tell you.

III. THE SOLE DEFLECTOR

We had no colonies on the Moon or Mars in 2025 at the start of the Athabasca Comet frenzy, but we had the technology to begin, including carbon nano-tube technology sufficient to build space elevators in those lower-gravity environments. Amazingly rapidly we were able to homestead on the Mars and Moon surface, and not long after that we put up orbiters around them. While these were being built, and most of Earth's population was digging into the ground to build cave habitats, some of our family was building the Deflector.

It seemed inevitable that the least amount of Earth's resources went into the construction of the Deflector, because it was the option with the least chance of success. So little was known about Athabasca's composition, that we could not have much certainty about what method might be most successful in deflecting it. Oddly, the Deflector was the most popular area of argument in the media. Proponents and opponents of the deflector, and advocates for various techniques, seemed to express their opinions with a melodramatic self-righteousness or even rabidity. A circus-like atmosphere arose around the debates over the Deflector, attracting a kind of freak-show and voyeur cult following. Debate about the need for either a strategy of exploding Athabasca, or deflecting it never arose, it seemed; the subject started with name-calling and soon devolved into sabotage and death threats. Meanwhile the work around the underground shelters and Orbiters progressed with a tone of ant-like methodicalness.

In the end, all efforts on attacking Athabasca directly settled on using an nuclear explosion, but still the rocket that would carry out this mission somehow came to be misnamed the Deflector. Hal Klay-Varney worked on the building of the Deflector, and during those years we hardly saw him because he was in 30 different countries coordinating all the pieces of the Deflector coming together under extreme secrecy. Because it was the sole Deflector, and because the project had been the subject of such controversy and even violence, the Deflector could not risk any more public exposure. At the last minute the pieces were flown to Kwajalein Island, and that picture there of your grandfather in front of that – that thing, is one of the few remaining photos of the Deflector.

Maybe people were right to fight so fiercely about the method of the Deflector, because in the end, even though your grandpa Hal fought to save the Earth, the choices his team made have resulted in the Earth's near-total decimation.

Athabasca could have been a mostly solid mass of metal, or rock, and things might have turned out better for the brave and the fearful masses on Earth huddled under miles of mountains. But Athabasca – it's one of the great tragedies of humanity – was a loose conglomeration of ice and small and massive fragments. So when Deflector implanted, and detonated inside of Comet 2036, much of it was redirected, but nine massive, smaller cometoids pummeled Earth in 2036, and we just call that time Brimstone.

Your grandpa Hal, God love him, was destroyed by the failure of the Deflector, and his refusal to take shelter in the underground cities was a kind of suicide.

Now before I can tell you about your Mommy and Daddy, I have to tell you something more about life and death and families in the orbital colonies.

IV. COMPLICATED FAMILIES

Back in the last great flowering of Earth culture, before Brimstone, most families were started by a man and a woman, and sometimes after couples had children they stayed together, and sometimes they didn't stay together – back then I called them "complicated families!"

Grandma Georgina laughed at this point. It made quite an impression on me because I had rarely heard her laugh, maybe never. This time, the laugh went from true amusement, to a kind of stressed laugh, as I remember it now. Then it turned into the raspy cough she had just started making.

Back then, some families had less than two children, and some families had more than two children. Can you imagine me growing up with seven brothers and sisters? That's that photo there of course, with all of us looking like a staircase of clones, but back then it was actually illegal to clone people! We were born as everybody was, after nine months in our mother's belly. A "natural childbirth" now means something totally different from what it did then!

Grandma smiled broadly again, but resisted the impulse to laugh, I guess because of the cough.

Your Mom, Deb, she was my only girl after Brimstone. I lost your aunt Helen in Brimstone, but I don't let myself feel sorry for myself because thousands of other people lost their lives and lost their loved ones in the failure of the Cubabi undergrounds. Maybe I'll tell you that story sometime!

Anyway your Mom, my little Debi, had gorgeous straight brown hair – straighter and browner than yours – and that's what your Daddy always said was her most beautiful feature. He used to joke about how he should have been a hair scientist, or how she could spend all the money she wanted on her hair...

Okay, honey, I know you're getting impatient and I'm getting a bit off-track, you've heard that part of the story before. But thank you for listening to what I have to tell you today.

Well okay, to the heart of the matter.

There was a long pause here, and Grandma Georgina drew a few labored breaths. Even though I was only about twelve, I was starting to realize that she was getting more frail every day. I knew not to interrupt Granny then, even if she wasn't speaking.

Well Davi, here's the truth. Your Mom and Dad didn't die when you were two. You had an older brother, and you were named after him. Mom and Dad, and your older brother Davie were on board the Arch Tachyon LifeBoat when it left Earth in 2033. I'm sorry to have to tell you this now Honey, that I lied to you all these years. I wanted to wait until you were old enough to understand, and you know I won't be around forever and I need to tell you the truth. You need to know your family history.

Davi, Honey, I missed them so much. Back when I was young... it was against my religion and my beliefs to... Davi, Honey, you're a clone of your older brother Davie. I hope you can forgive me for keeping the truth from you all these years. I'm so glad to tell you the truth now.

Georgina breathed deeply and more easily now than she had before. Her eyes looked softer to me.

So Davi, I love you so much and it doesn't matter to me what I thought back on Earth or... well I think when you're older you'll understand better... well Honey, that's what I needed you to know so bad I've kept you here all morning with my stories.

Well, yes Granny that was a long story, but you know I love your stories.

Well Davi, go ahead, I know that's a lot, and you must have some questions to ask me. Go ahead ... do you want to ask me any questions?

There was a pause of about five seconds.

Yes, Granny. Can I go out now and play with Mena and Mela in the microgravity lab?

•••

DAN SHAW, B.S.

DanShaw.com lives in Vortex, Virginia with his second ex-wife and their two toy poodles, Goth and Grunge.

Just kidding.

Dan is a filmmaker, webmaster of BeyondEarth.org, and mostly writes poetry and non-fiction. He is studying collaborative techniques and technologies in aerospace. Please visit DanShaw.com.

SHERRY BELL, PH.D.

Dr. Sherry Bell received her Bachelor of Science degree from California State University, Stanislaus and her Ph.D. from Capella University. She is an Industrial/Organizational Psychologist. Sherry specializes in providing personality-based services. Her services include executive coaching, goal setting and goal achievement, vocational guidance, and weight management. She is an avid researcher and has been awarded two research grants.

Sherry is a member of the American Psychological Association and the Society for Industrial Organizational Psychology. She is a lifetime member of both the Golden Key National Honor Society and Psi Chi (the National Honor Society of Psychology). She is an active member of the National Space Society and has for the past two years chaired a Track at the annual International Space Development Conference.

Her recent accomplishments include:

(1) Sherry is a Senior Leadership Advisor for the Aerospace Technology Working Group (ATWG).
(2) She is a published author, and has given numerous presentations.
(3) Sherry has been a guest on a Space-based radio program.

Dr. Bell can be contacted at: DrSherryBell@aol.com.

CONCLUSION

CHAPTER 20

CONCLUSION FICTION

THE COLONIES, 2084

SHERRY E. BELL, PH.D.

I lovingly crafted this reverie as a tribute to each of the authors who contributed to this book. In this science fiction story, I incorporated something from every chapter in this book, Living In Space: Cultural and Social Dynamics, Opportunities and Challenges In Permanent Space Habitats. I also paid tribute to our first ATWG sponsored book, Beyond Earth: The Future of Humans in Space.

In this chapter I offer the reader sex in the form of sex and reproduction, drugs in the form of pharmaceuticals and wine, and rock and roll in the form of music in Space.

•••

INTRODUCTION

Who:

From: Dr. Sherry Bell, Administrator of the Mars Colony

To: Interested parties

What: Report on the Colonies

When: November 26, 2084

Where: This message originates from the Headquarters on the Mars
Colony

How: This message is being transmitted via Meta-web

Why: In response to your requests

HOW THE COLONIES CAME TO BE

In 2025, astronomers on Earth discovered a comet, named Comet ME-
2037, which was on a collision course with Earth. Were the comet to hit
the Earth, mass extinctions would be the result and human life as we know
it, would likely vanish. In an unprecedented global cooperative effort, all
the resources of the world were pooled into one common effort - the
survival of Humans.

In 2015, the U.S.A. had established a Space Commerce Agency (See
Hsu and Cox, Chapter 15). Its purpose was to oversee all of nation's
commercial Space activities such as tourism as well as the exploration
efforts conducted primarily by NASA. After comet ME-2037 was
discovered, the Space Commerce Agency became the busiest agency in the
world.

In the hopes that one would prove to be the lifeboat on which
humankind could pin its hopes, dreams, and even its future, a number of
strategies were implemented. The people of Earth decided to rapidly
construct a variety of colonies. Although no permanent colonies in space
had yet been established, the technology was available to build them.
Colonies were established on the Moon and on Mars. Space elevators from
the surface of each of these were constructed. Orbiting Stations were built
and placed in the orbits of both the Moon and Mars. Four Tachyon Space-
faring ships were constructed and launched.

Building an anti-matter vessel was considered; however, the cost was
too high. All of the Earth's economic resources would have had to be spent
on one project. No other projects could have been considered.

Because we were trying to get as many of us as we could off of the
surface of Earth within a constrained time-frame; we chose not to put "all
our eggs into one basket." We constructed small capsules containing a
representation of the DNA of all Earth's bio-organisms, including Humans,
and sent them out into Space in all directions.

By the year 2035, all of these projects were completed. In fact, one of the Tachyon ships that had been launched in 2033 returned to Earth six months later. They had not made contact with any alien life during the six months they were gone. Well that is what we thought at first. Quite soon it became apparent E. T. life had returned with them in the form of microorganisms (See Bell, Chapter 13). In the year 2034, it looked like big comet ME-2037 was not going to be the thing that wiped out humanity; instead it would be small virulent microorganisms. Fortunately scientists found a way to eradicate the disease caused by these deadly little critters that had been brought back – but only after more than 525 million people had died.

Another of the ships made contact with alien intelligence. They sent a message back to Earth which read, "Contact has been made. The quintessence of reality is the melding of perception, context, and the ineffable," (See Hannon, Chapter 18). We received the message; however the ship did not return and we have not heard from them again.

The valiant efforts to deflect the Comet ME-2037 resulted in it breaking into a dozen large pieces and countless smaller ones. Nine of the large pieces impacted with Earth in 2037. Much of the life on the planet was destroyed; however a few Humans in isolated pockets survived.

The Colonies Today

Sex and Reproduction

Having sex is encouraged on all the colonies. It is widely recognized that the release of sexual tension results in a positive state of mind. Although the neurochemical basis of this is not yet well-understood; the positive benefits of having sex are plain to see.

Within the colonies, copulation between people over the age of 16 is permitted; however no female on the Mars Colony can bear live young. All embryos are grown in our exogenesis laboratory where they are subjected to genetic testing, and their unique DNA is copied and stored. Initially, males raised a great stink about this; however they began to embrace this reproductive strategy after two women died while pregnant - as had been foretold by Strongin and Reese in 2009 (See Chapter 2).

"Unless pregnant women lived in zero gravity chambers or large bodies of water, the compromised pelvic bone integrity would make carrying a child to term difficult, even with Mars' lower gravity. Natural childbirth would be prohibitive" (Para. 2 in Sexuality and Reproduction Section).

All colonies are provided with exogenesis chambers. Eggs are harvested from females and fertilized by sperm from males. All offspring are genetically modified to make them better suited to the environment in which they are expected to thrive. After 40-plus years men have come to realize mothers are still as loving and devoted as they were before - when they bore the live young.

GENETIC MODIFICATION

The offspring of certain groups are genetically modified in order to help them better adapt to the environment in which they live. Humans on Earth are not genetically modified except by the very long process of natural selection. Earth is a perfect place for the unmodified Human body as it is.

Humans who boarded the Space-faring ships were not modified, however the technology and equipment necessary to implement the changes were sent on-board the vessels. It was not known what changes would be evolutionarily adaptive for them. Nor is it known which modifications, if any, each of them has applied.

The Orbiters, who live in zero gravity, modified their offspring to have two arms and no legs. Some of them have been modified to have four arms and no legs. Legs are a nuisance in zero g.

One thing each of us off-Earth has in common is that we have been genetically modified to help us adapt to our respective environments. We have been modified to resist radiation and to address bone loss and muscle atrophy issues among other things. Although we do not directly produce live offspring, we can become parents if we choose.

The genetic modifications allow us to live much longer than before. It is conceivable we might live as long as 500 years. By then technological advances will have made even this figure obsolete.

PAIR BONDS

Monogamy, once the most common form of mating structure on Earth, is not encouraged on any of the colonies. With the removal of the fear of unplanned pregnancies and all the benefits that are known regarding sexual intercourse, there is no reason not to pursue multiple partners if desired.

Unfortunately, we did not come to this conclusion until after there had been several homicides between romantic rivals. At least one occurrence happened on each Mars, the Moon, the Morris Orbitor, the Cox Orbitor, and even one on the International Space Station. The tradition of monogamy died hard. We, on Mars, hope that someday in the near future sexual and romantic jealousy will be eradicated, and with it, the attendant violence.

THE SETTLERS

Astrosociologists predicted our societies would be well-served were we to give lots of thought and attention to the matter beforehand (See Pass, Chapter 7). They decided to work with psychologists. Together those two groups determined "best-fit" models. For the most part, they got things right. Much like in Eckelkamp's fiction (See Chapter 17) we respect one another and live in harmony. All people are cherished - young and old alike (See Wong, Chapter 8).

MUSIC AND ART

The second in command here is Dr. Gianna Pittman. She is an artist of Solar-System re-known (See Brzezinski, Chapter 16). She is both a visual artist and a musician. Because she was afraid they would be destroyed by the comet and might become lost to us forever, she brought as many of the "Pythagorean Lambdoma Harmonic Keyboard" (PLHK) instruments with her as she could (See Hero, Chapter 6). Barbara Hero, the inventor of this instrument, described the experience of playing the PLHK in the following manner.

It becomes a kinesthetic experience that can induce a pleasantly altered state, one where there is no anger, only love. I have observed that for many people with problems to solve on the emotional, physical, spiritual or mental levels, solutions often come instantaneously. The instrument leads to a peaceful, harmonic way of life, as envisioned by Pythagoras in 500 BC, when he strummed his lyre and dissipated anger in one of his disciples (Para. 3 in Conclusion Section).

The other colonists are immensely grateful to Dr. Pittman for bringing these with her. Without her foresight, they would be without this miraculous healing instrument.

HABITATS, GOVERNANCES AND OBJECTIVES

HABITATS

Here on Mars we live underground. So do most people on the Earth and Moon. This allows us to be much better protected against the elements than we would be if we were living on the surface.

Every person, including every child, is provided with his or her own private quarters. The effects privacy has on morale and sense of well-being--as well as general productiveness have long been known (See Ryan and Kutschera, Chapter 9).

The first Humans on Mars were miners. They lived in inflatable habitats. Many of those original habitats are still in use today. Dr. Gianna Pittman uses one as an art studio and another for her music studio (See Brzezinski, Chapter 16).

After two years of tireless effort, the miners completed the underground habitat we call home (See Taylor and Benaroya, Chapter 10). We owe our existence to those early miners and to honor them, we declared the site where they first erected their inflatable habitats a historic place. In addition we turned one of the structures into a museum. We named this museum the "Sadeh Museum" to honor the two men who designed the buildings (See Sadeh and Sadeh, Chapter 14).

Most of the people on Earth live underground in one gigantic manmade cave that was constructed decades before Comet ME – 2037 struck. The area where they live is located in Colorado in the U.S.A. and was formerly the headquarters for an organization called NORAD. Before the impact, the underground facility was used to watch for incoming missiles or aircraft attack. It is pretty much missile-proof. Thankfully the Space Based Solar Power system that had been put in place earlier in the century remained intact after the impact. The system provides the people with ample clean energy (See Hsu, Chapter 12).

The Morris and Cox Orbitors are breathtaking to behold, and appear much like Paolo Soleri (See Chapter 4) envisioned they would. Soleri described space cities as being "luminescent, chromatic, and serene," which is indeed how they appear.

GOVERNANCE

Before our colonies were settled we decided each colony would be independent and would have its own government. When drafting our respective constitutions, each of us incorporated elements from Bob Krone's Overview Space Governance model (See Chapter 3). We also drew heavily from his "A Code of Ethics for Humans in Space" (Connor, Downing and Krone, 2006). Each colony adapted Krone et al.'s earlier works to suit their particular requirements.

OBJECTIVES

Here on Mars our primary objective is to maintain our habitat. Mars can be a hostile place to live if we are not careful. Our first priority is to ensure that our food, water, and air supplies remain stable.

Because we cannot depend on others for food, and in fact the people on the Cox Orbiter depend upon us for this commodity, growing enough food is essential to our survival. We do this by growing food in our greenhouses. We have acres and acres of greenhouses on the surface. Nothing is wasted. Inedible plant matter is used to produce fuel (See Kiker, Chapter 11). A third objective is to terraform our planet.

We have discovered that our people are more content and our colony is enhanced when individuals have opportunities to master different trades. An example of how this has worked well follows. A few years ago one of our mechanical engineers decided he would like to work on the surface of the planet for a change. He elected to work in the greenhouses. He found he felt happiest when working in what was then a small out of the way vineyard. It did not take him long to pair his mechanical engineering talents with his love of working with fruit-bearing vines. Before long we had wine -- which we use in large part for ceremonial purposes. Word of that invention spread quickly, and now the wines we produce are highly coveted by the other Colonies.

One of the goals for the colonists living on the Morris and Cox Orbiters is to conduct experiments. Another goal is to produce medications. Due to the heavy biology research focus that takes place on the Orbiters, most of the medical expertise comes from them.

The primary mission of the Moon Colony is to build and launch Tachyon ships. We all remain hopeful and optimistic about the successes of these endeavors.

WE HAVE BECOME A CIVILIZATION OF PEACE

The Colonies including Earth have collectively become a civilization of peace. We experience "...the peace of Shalom, the peace of civilization..." (Kirby, 2008).

When establishing our colonies, we were on the offensive. Our aim was to save the Human race. We did not spend our last days fighting one another or making instruments of war. Dan Shaw expressed this nicely when he wrote,

"Countries came together against a common enemy, and no resources were destroyed or squandered in wars and deadly conflicts." (See Shaw and Bell, Chapter 19).

We feel a sense of interconnection. Because our systems are even more inter-related than they were in 2009 when Lowry Burgess (See Chapter 5) so eloquently described the Metasphere, the Metasphere is still present. However, perhaps because none of us has any military-type defenses, or maybe in part due to our collective close brush with death, this Metasphere seems much more benign than the one described by Burgess in 2009. Today's Metasphere now encompasses the Earth and all of the Colonies combined.

We experience what Frank White described as "a new form of unity," (See White, Chapter 1). We know, see, feel, and intuit each other. We have discovered Frank White was right when he said, "The universe, or Cosma, is, like the Earth, an interconnected and interdependent system" (in Section "The Cosma Hypothesis").

Humans now see the Cosma in all its splendor and glory!

•••

SHERRY BELL, PH.D.

Dr. Sherry Bell received her Bachelor of Science degree from California State University, Stanislaus and her Ph.D. from Capella University. She is an Industrial/Organizational Psychologist. Sherry specializes in providing personality-based services. Her services include executive coaching, goal setting and goal achievement, vocational guidance, and weight management. She is an avid researcher and has been awarded two research grants.

Sherry is a member of the American Psychological Association and the Society for Industrial Organizational Psychology. She is a lifetime member of both the Golden Key National Honor Society and Psi Chi (the National Honor Society of Psychology). She is an active member of the National Space Society and has for the past two years chaired a Track at the annual International Space Development Conference.

Her recent accomplishments include:

(1) Sherry is a Senior Leadership Advisor for the Aerospace Technology Working Group (ATWG).
(2) She is a published author, and has given numerous presentations.
(3) Sherry has been a guest on a Space-based radio program.

Dr. Bell can be contacted at: DrSherryBell@aol.com.

REFERENCES

1. Connor, K. T., Downing, L., & Krone, B. (2006). "A Code of Ethics for Humans in Space." In B. Krone (Ed.), *Beyond Earth: The Future of Humans in Space*, (pp. 119-126). Burlington, Ontario, Canada: Apogee Space Press.
2. Kirby, R. (2008, November). "The once and future Veterans Administration." Unpublished manuscript, Kepler Space University, Seattle, WA.

CHAPTER 21

CONCLUSION
NON-FICTION

LANGDON MORRIS

In the course of our day-to-day lives in Earth's industrialized nations, we tend to take for granted the enormous complexity of the global civilization upon which we thoroughly depend. Every day the food that we eat, the coffee we drink, the cars we drive, and indeed a great many of the products as well as the information that we consume, wear, use, and enjoy, all were grown, prepared, created, and/or manufactured hundreds or thousands of miles from where we live.

They were produced by human hands, robotic arms, computers, and massive factories, and brought to us via amazing supply chains that coordinate the activities of hundreds of thousands of people with extraordinary grace, timing, and precision. Likewise, we consume sounds, images, and text that was created thousands of miles away, and transported to us via energy and communications systems that span the globe and reach far into space. And without our space-based GPS navigation systems, we get lost on the way to the grocery store.

We participate vicariously in global forums for decision making and governance via elections in which we choose government officials and corporate boards of directors on every continent, who then make decisions about war, peace, economics, business, and culture that affect the lives of all 6+ billions humans, and trillions of other living creatures with whom we share this planet.

And now, as humans make the transition into space, the extraordinary complexity of our civilization becomes apparent to us in new ways as we consider how the complexity that we take for granted on Earth must be painstakingly reproduced off Earth in the new social, physical, and informational context of colonies, cities, and nations that we believe we will one day build on the Moon, Mars, in orbit, and beyond.

And of course, these off-world worlds will inevitably grapple with yet another dimension of complexity that on-Earthers won't have to deal with, providing a life-sustaining atmosphere.

But why would we do this? What go to all that trouble? For the thrill? Yes, many will do it exactly for that reason. For the challenge? Yes, that too. And for our survival? Yes, perhaps one day we will discover that human survival does indeed depend on our ability to adapt human culture to other planets, moons, orbiting stations, and space-faring craft.

•••

This book, *Living in Space*, is sponsored by The Aerospace Technology Working Group, an extraordinary collection of veterans and newcomers to the as-yet nascent space age. We believe that humans are indeed destined to one day venture into space in long-lived (i.e., permanent) habitats, and whether we call them colonies, worlds, orbiters, or cities is less important than the fact of our intention to venture forth beyond earth. Indeed, Beyond Earth was the title of our first collection of writings, published in 2006. Now, in this, our second book, you have read the contributions of 25 authors who examined many additional aspects of this challenging and enticing future that awaits us.

If we look at their writings in aggregate we begin to appreciate the enormity of the task that we face. The contributors to this volume examined psychology, sociology, physiology, the visual arts, music, health, medicine, reproduction, governance, economics, management, energy, construction, information, microbes, colonization, joy, beauty, fear, and love, all certain to be critical elements of the quest to survive and thrive off of our home planet, just as they are now critical to our survival on the home planet.

Some of the authors contributed essays on technical topics or philosophical conjectures, while others wrote stories to help us envision our future lives on future worlds. As we come to understand what it means to live in space, we recognize that both fiction and non-fiction contribute to our understanding of life on and off-world, and to our appreciation of what it means both to be human, and to become still more fully human.

And while these ideas occupy the thoughts of many of us in the globally-interconnected, industrialized world, it's not just the Earth's wealthiest individuals and nations who appreciate the significance of the space enterprise for human culture and for the global and now meta-global economy.

Today, the emerging powers of China and India already have ambitious efforts under way in space exploration and development, and they have deployed extraordinarily talented and dedicated scientists, engineers, scholars, and designers in pursuit of their own space aspirations. Because these are not only national aspirations, but in fact the quest for space is a human aspiration.

For millions of years our ancestors gazed in awe at the stars, just as we still do today. They wondered what lies beyond, just as we still do today. And now, on a foundation of mighty technology and inspiring willpower and ambition, we have begun to reach out, and soon we hope to touch what we once only dreamed about. Already there are permanent residents on the International Space Station; how long will it be until there are entire cities in space?

At this threshold of the space age, then, we may perhaps be on the verge of enriching our knowledge, our cultures, and indeed all of human civilization with answers to some of these questions, answers that are sure to surprise, astound, change, and perhaps even frighten us.

Our thanks to you, the reader, for sharing this journey with us.

•••

LANGDON MORRIS

Langdon Morris is a partner of InnovationLabs (www.innovationlabs.com), one of the world's leading innovation consultancies. He is recognized globally as a leader in the field of innovation, and his recent clients include organizations such as NASA, GE, Gemalto, Total Oil, the Federal Reserve Bank of the US, Johnson & Johnson, Tata Group, France Telecom, Stanford University, Cap Gemini, Wipro, L'Oréal, Accor Hotels, and many others.

He is a member of the leadership team of The Aerospace Technology Working Group (ATWG), and also Senior Practice Scholar of the Ackoff Center at the University of Pennsylvania where he is researching complex social and business systems. He is a Senior Fellow of the Economic Opportunities Program of the Aspen Institute, a member of the Editorial Committee of the International Journal of Innovation Science, and a member of the Scientific Committee of Business Digest, Paris.

He has taught MBA courses in strategy at the Ecole Nationale des Ponts et Chaussées in Paris and Universidad de Belgrano in Buenos Aires.

He is the author or co-author of numerous white papers and highly acclaimed books on business (4 books), and US politics (2 books), with editions in Japanese, Chinese, Korean, and French. He was a contributor to and co-editor of the previous book in this series, *Beyond Earth: The Future of Humans in Space*. He is highly sought after as a speaker and workshop leader, and participates frequently at conferences and workshops worldwide.

INDEX

A

B

LIVING IN SPACE
IS A PUBLICATION OF
THE AEROSPACE TECHNOLOGY WORKING GROUP

WWW.ATWG.ORG